CROP PRODUCTION IN WATERLOGGED SALINE SOILS

Related Books

Title	*Author*
• A Manual on Conservation of Soil and Water	*USDA*
• Agricultural Salinity Assessment and Management	*K.K. Tanji*
• Agriculture and Fertilizers	*O.C. Bockman*
• Agrochemicals in Plant Disease Management	*N.G. Ravichandran*
• Biofertilizer Manual**	*FNCA*
• Biofumigation and Solarization for Management of Soil-Borne Plant Pathogens	*P.P. Reddy*
• Comprehensive Assessment of Soil Health *3rd Ed***	*SIPS*
• Crop Production in Salt Affected Soils	*I.C. Gupta*
• Crop Production in Waterlogged Saline Soils	*S.K. Gupta*
• Crop Protection Strategies under Climate Change Scenarios	*P.P. Reddy*
• Drainage Engineering: Principles and Practices	*S.K. Gupta*
• Fertilizer Manual	*IFDC*
• Fundamentals of Soil Science	*A. Rathinasamy*
• Fundamentals of Soil Science**	*R.L. Arya*
• Handbook of Saline and Alkali Soils Diagnosis Reclamation and Management**	*S.K. Gupta*
• Managing Manure**	*Mark Kopecky*
• Manual on Fertilizer Distribution**	*FAO*
• Methods of Analysis for Soils of Arid and Semi Arid Regions**	*I.I. Bashour*
• Methods of Soil Analysis: Microbiological and Biochemical Properties (Part 2)**	*R.W. Weaver*
• Nanofertilizers: Challenges and Prospects **	*J.C. Tarafdar*
• Nanotechnology	*J.C. Tarafdar*
• Organic Farming Theory & Practice	*S.P. Palaniappan*
• Salinity Tolerance in Plants: Methods, Mechanisms and Management *2nd Ed*	*B.K. Garg*
• Salt Affected Soils: Reclamation and Management	*S.K. Gupta*
• Soil and Plant Analysis	*C.S. Piper*
• Soil and Water Conservation in Semi-Arid Areas	*N.W. Hudson*
• Soil Chemical Analysis	*M.L. Jackson*
• Soil Conservation Technical Handbook	*D.H. Hicks*
• Soil Guide: A Handbook for Understanding and Managing Agricultural Soils	*Geoff Moore*
• Statistical Methods in Water**	*D.R. Helsel*
• The Drought Resilient Farm**	*Dale Strickler*
• The Soil-Plant System in Relation to Inorganic Nutrition	*M. Fried*
• The Use of Saline Waters for Crop Production	*FAO*
• Water Measurement Manual *3rd Ed*	*USDI*

CROP PRODUCTION IN WATERLOGGED SALINE SOILS

S.K. Gupta
Central Soil Salinity Research Institute
Karnal, 132 001

I.C. Gupta
Central Arid Zone Research Institute
Jodhpur - 342 003

SCIENTIFIC PUBLISHERS

Published by:
Scientific Publishers
5A, New Pali Road, P.O. Box 91
Jodhpur 342 001, India
E-mail: info@scientificpub.com
Website: www.scientificpub.com

Print : 2025

ISBN: 978-81-72331-59-7
eISBN: 978-93-88449-48-9

Printed in India

PREFACE

To even out the inequity in the distribution of water resources in the country and to meet the water demand for domestic, industrial and agricultural sectors of the arid and semi-arid region, India decided to develop its water resources. The approach called for water conservation, storage, diversion and transportation of water from regions with excess water to the deficit ones. Such an approach initially led to increased production and productivity and provided needed food security to the country. Nonetheless, the approach has now come under attack from many quarters as a result of environmental degradation. The most conspicuous adverse environmental effects have appeared in the form of waterlogging and soil salinity affecting land and water resources. The vast stretches of once productive agricultural lands have gone out of cultivation leading to the development of "wet deserts".

The problem of waterlogging and soil salinity is not only confined to irrigated land but land in coastal humid and sub-humid regions is also affected by the twin problems due to tidal influence or some other action. It has been estimated that overall about 5.1 million ha of agricultural land is suffering as a result of waterlogging and/or soil salinity.

Researchers in the country have come out with sound technically and economically feasible technologies for the reclamation and management of waterlogged salt affected lands. The technologies for the reclamation and management of alkali lands have been well received. A large chunk of these barren lands have been reclaimed which is now producing food grains valued at Rs. 1000 crores per annum. However, the technical breakthrough achieved in the reclamation and management of waterlogged saline soils has not yet been translated into field practices at the farmers' field. It is recognized that it could be as a result of non-availability of these technologies to the people in the affected areas and as a result of technologies being costly and requiring a group approach rather than a single farmer approach which was applicable in the case of alkali land reclamation.

Realizing these handicaps, an attempt has been made by

the authors to collate the existing technologies and put them in print in a manner that these could be grasped by the concerned people. The book has been logically divided into chapters, successively dealing with the technological components in each chapter. The first chapter discusses the extent, distribution and nature of the problems to familiarise the readers with various issues. The diagnostic analysis procedures are discussed in chapter 2. The issue of soil salinity in the root zone which is the root cause of non-sustainability of irrigated agriculture is discussed in chapter 3. Technologies for water table control which could be applied both in humid/sub-humid and semi-arid/arid regions is discussed in chapter 4. The technology for leaching of saline soils and related issues form the part of the 5th chapter. Crop production practices and the alternate land management technology are included in chapter 6. Reclamation of coastal saline soils which in itself is specialized task is discussed in a separate chapter. Most of the issues that have been discussed for waterlogged inland saline soils have been briefly discussed in this chapter for the coastal regions. Differences in the approach for these soils vis-a-vis coastal saline soils have been brought out. Finally the socio-economic aspects which are important to decide the economic viability of rehabilitation projects have been included in the last chapter. Unlike in other books where nomograph and tables are included for drainage design, herein computer programmes have been included which is the modern approach in dealing with issues of design and development.

In this book, the work of many scientists and institutions have been included. Herein, we acknowledge the contribution of each of them. We are specially thankful to Dr. NT Singh, Dr. KVGK Rao, Dr. OP Singh, Dr. DP Sharma, Dr. Gurbachan Singh, Dr. KN Singh, Dr. AK Batra, Dr. A. Swarup and others of the Central Soil Salinity Research Institute, Karnal who have been in the forefront of developing reclamation technologies for the reclamation and management of such degraded lands.

We hope that this book will be immensely useful to researchers, teachers and students, command area authorities, extension workers and above all the farming community of India. Any attempt made in the reclamation with the help of this book will give us immense satisfaction.

— Authors

CONTENTS

1

EXTENT, DISTRIBUTION AND NATURE OF PROBLEM SOILS

SOIL RESOURCES OF INDIA

The land mass which forms the subcontinent of India, covers an area of 329 mha. Soil survey reports of India have identified more than 26 soil types. Raychaudhari (1962) quantitatively estimated the area under different soil groups (Table 1.1). It is observed that the alluvial soils, the black soils, the red soils and the laterite and lateritic soils are the major soil groups of significance.

Alluvial soils is by far the largest and the most important soil group spread over the Indo-Gangetic plains, the Brahmaputra valley and deltas of several major rivers. Newer alluviums are sandy and have less calcium carbonate content and are lighter in colour. Older alluviums are more clayey in composition and have a layer of $CaCO_3$ concretion at shallow depth (1.0-1.5 m). Nitrogen deficiency is exhibited to a high degree in most of the alluvial soils and is a major limiting factor in crop production. In saline-alkali soils high exchangeable sodium and/or high salt content also limits crop production.

Black soils are the typical soils derived from the deccan trap as a result of weathering of a variety of parent materials. Deep black soils derived from trap basalt, gneiss, shales, schists and limestone are often of mixed origin. Shallow black soils are lighter in colour and have lime layer exposed to the surface. The black soils are fairly fine textured with clay content between 35 and 60 per cent. These soils contain moderate amount of salts. It is likely that these soils may turn saline/ alkali upon introduction of irrigation.

Table 1.1 Area under different soil types

Type of soils	Approximate area (mha)
Alluvial soils	101.2
Alluvial soils highly calcareous	8.9
Coastal alluvium	8.5
Deltaic alluvium (occasionally saline)	17.0
Alluvial soils affected by salinity and alkalinity	6.9
Desert soils	14.6
Deep-black soils	6.9
Medium-black soils	18.6
Shallow-black soils	4.9
Black soils affected by salinity and alkalinity	6.9
Black soils undifferentiated	12.5
Mixed red and black soils	10.5
Red soils	30.4
Red gravelly soils	1.6
Red and yellow soils	17.8
Laterite soils	10.1
Laterite and lateritic soils	2.0
Brown soils under deciduous	1.6
Grey and brown soils	3.6
Hill soils	2.4
Podsolic soils	3.6
Forest soil laterised	6.5
Foothill *tarai* hill soils	5.7
Mountain meadow soils	11.7
Skeletal soils	2.4
Peat muck, very humus and humus soils also called bog soils, organic soils and half-bog soils	0.2
Other soils (viz. mountain soils undifferentiated glaciers and eternal snow)	11.8
Total	328.8

The ancient crystalline and metamorphic rocks on meteoric weathering have given rise to what is called red soil. These soils are derived from granite, gneiss sandstone, basalt and shales and have originated from weathering of crystalline and metamorphic and other sub-ordinate rocks. These are rich in ferro-magnesium minerals and have kaolinite predominance. A diagnostic argillic horizon develops in these soils as a result of clay transport. Depending upon the vegetation and topography, red soils differ in texture and colour.

Laterite and lateritic soils are formations peculiar to India and some other tropical countries. It is a compact to vesicular rock composed essentially of a mixture of the hydrated oxides of aluminum and ions with small amounts of manganese oxide. All laterite soils are poor in lime and magnesia and deficient in nitrogen. Occasionally the phosphate content may be high.

There are many other groups of soils which also occupy extensive areas in the country, however, details about these soils can be found in Soils of India and Their Management (FAI, 1985) and Benchmark Soils of India (Murthy *et al.*, 1982).

Problems Associated with Land Resource

Land is the basic resource base of production in agriculture, forestry, animal husbandry etc. While demand for food, fibre and fodder is increasing as a result of growing human and livestock population, the vital soil resource is almost inelastic. The net sown area of about 119 mha in 1950 expanded to 143 mha in 1980 and is likely to go up to 155 mha in 2025. With increasing demand from non-agriculture sectors, such as residential, industrial and areas for roads, railways and irrigation/drainage systems, it is most likely that area for horizontal expansion might not be available. Thus, appropriate means to increase agricultural production would hinge upon our efforts to manage the existing soil and water resources to obtain maximum production.

Presently available land resource of the country is afflicted with a number of problems causing overall poor yields (Table 1.2). It is revealed that the problem of salt affected soils is serious on account of its extent. Degree of damage is also serious once the land resource is infested with waterlogging or soil salinity. Although, salt affected soils are potentially productive, but these have gone out of cultivation because of presence of excess amounts of salts. The most worrisome feature is that the land area under waterlogging and soil salinity is increasing over the years. In this background, scientific reclamation of the salt affected soils, post reclamation management and prevention of the spread of formation of alkali/salinity to newer areas assumes greater significance.

Table 1.2 Extent of problems of land degradation in India

Nature of problem	Area affected (mha)	Percentage of total area
Geographical area	329	100
Area subject to water and wind erosion	168.2	51.1
Area degraded through special problems		
Waterlogged area	16.3	4.9
Salinisation/ alkalisation including nutrient deficiency	14.8	4.5

WATER RESOURCES OF INDIA

The country as a whole receives an annual rainfall of about 120 cm which over the total land area of 329 mha yields an annual precipitation of 400 mha-m. A percentage of this rainfall is lost as run off while some is recharged to the ground water aquifer.

Surface Water Resources

For estimating the surface water resources, the country has been identified into six river basins, namely Indus basin, Ganga basin, Brahmaputra basin, East Coast, West Coast and the Rajputana region. The First Irrigation Commission (1903) estimated surface water resources of the country as 144.32 mha-m. It was later on revised to 167.23 mha-m by Khosla (1949). Statistical estimates of the Central Water and Power Commission put this figure at 188.21 mha-m. A comprehensive assessment made by the National Commission on Agriculture (1976) indicated that surface water resources of the country are 178 mha- m. According to the present indications, the total and utilisable surface water resources of the country are 176.8 m ha-m and 70 mha-m respectively. A upward revision in the utilisable flows may be expected on account of advancement in technology and on account of large scale inter-basin transfer of waters.

Ground Water Resources

Raghava Rao *et al.* (1969) estimated that about 42.6 mha-m of replenishable ground water resources are available in the country. This estimate has been endorsed by the Task Force on Ground water resources of the Planning Commission. Considering losses from evapotranspiration, subsurface flows, recharge in the saline water regions and efficiency etc., the Task Force recommended that on a long term basis, a potential of 20.36 mha-m of replenishable ground water resource exists for exploitation. This does not include ground water in storage. According to the Central Ground Water Board (Government of India), the country has a exploitable ground water potential of 32.3 mha-m.

A major limitation for the use of ground water has been its quality. As innumerable factors affect the ground water quality, wide variations in water quality can be observed within the country and within a state. The ground water quality survey has indicated that the poor quality waters constitute a major portion (32.84 per cent) of the total utilizable potential of ground water in arid and semi-arid regions (Table 1.3).

Utilisable Flows and Irrigation Potential

The utilisable flows aggregate to about 105 mha-m out of which 70 mha-m are from surface and 35 mha-m from the ground water resource. It may be understood that not all the water will be available for irrigation as there are other demands on it. It is expected that in 2025 AD, about 77 mha-m of water would be available for irrigation and the rest would be for domestic and industrial uses. The net sown area in 2025 AD is estimated to be 155 mha. With expected cropping intensity of 136 per cent, the total cropped area would be about 210 mha. The available water resources even at that time would be sufficient to provide irrigation to an area of about 110 mha. Therefore, with limited resources, it is expected that only 52 per cent of the gross sown area of 210 mha will be under irrigation in 2025. The remaining area will continue to depend upon the rainfall.

To usher in an era of self sufficiency in food production, reliance was largely upon the development and use of our water resources. In pursuance of this policy, 246 major, 1039 medium and 1.08 million minor irrigation projects to harness surface water were taken up between 1951 and 1990. In addition, during the same period, about 25.88 million ground water wells, were installed, raising the irrigation potential from 20.9 mha in 1951 to 71.8 mha in 1989-90 (Ministry of Water Resources, 1990). However, this achievement was not unblemished. In the wake of increased productivity and improved quality of life, there were also some adverse environmental impacts. Waterlogging and soil salinisation of agricultural lands is one of the most serious environmental problem having global dimensions.

Table 1.3 Ground water potential and percent distribution of marginal and poor quality ground waters in the arid and semi-arid regions of India

State	Utilizable potential (mha-m/yr)	Net draft (mha-m/yr)	Distribution of ground water (%)		
			Good	Marginal	Poor
Andhra Pradesh	3.66	0.74	68	25	7
Gujarat	2.03	0.69	70	20	10
Haryana	0.88	0.61	37	8	55
Karnataka	1.30	0.18	62	12	26
Madhya Pradesh	5.95	0.49	75	10	15
Punjab	1.31	0.93	59	22	19
Rajasthan	1.83	0.43	16	16	68
Tamilnadu	2.69	0.99	*	*	*
Uttar Pradesh	9.27	2.68	32	22	46
Total	28.92	7.74	419	135	246

* Not estimated

HISTORY OF IRRIGATION AND WATERLOGGING

Irrigation is an art which has been practised in India since times immemorial. In *Narada Smrity*, it is mentioned that *Rishi Narada* inquired from emperor Yudhisthra whether the farmers were sturdy and prosperous and whether the dams had water for distribution in different parts of the kingdom. The remains of Indus Valley Civilization and Vedic texts bear testimony to the practice of irrigation. The Grand *Anicut* across river Cauvery in Tamilnadu was constructed in the second century A.D. Feroze Shah Tuglaq in fourteenth century A.D. got the Western Yamuna Canal constructed. Irrigation works such as construction of a large number of wells all over the country, tanks in south India and inundation canals in north India were constructed during early nineteenth century. The total irrigated area through public works in 1900 stood at 13.3 mha which increased to 19.3 mha in 1920-21. At the time of partition, the total irrigated area stood at 28.2 mha out of which 8.8 mha went to Pakistan leaving 19.4 mha with India. Before the plan period started in 1951, the irrigated area had already increased to 22.6 mha. The irrigated area rose to 73.1 mha in 1991-92 (Table 1.4). During the eighth five year plan an irrigation potential of 15.8 mha is proposed to be created. It is however, a matter of concern that the irrigation potential created has not been fully utilized. The gap between development and utilization has been estimated to be 8.1 mha in 1991-92.

Inter-basin transfer of water to irrigate lands in arid and semi-arid regions interferes with the hydrological cycle. It creates imbalance between the recharge to the ground water and the discharge through the natural out flow. As a result water is taken into storage leading to rise in the water table. It is, therefore, felt that introduction of irrigation and development of waterlogging and soil salinity is closely interlinked. "Turning white of fields" has been mentioned in a Babylonian story of Atrasis (1800 B.C). In the Indian context, harmful effects of excess water are enunciated in the epic *Narada Smrity* where sage Narada says, "No grain is ever produced without water, but too much water tends to spoil the grain". An inundation is as injurious to growth as dearth of water. Practically, however, there is no mention of the problem till 19th century when the problems of waterlogging and soil salinity attracted the attention of the Government.

Western Yamuna canal, constructed in 14th century, was realigned by the British and opened for irrigation in 1839. The complaints of soil salinisation first appeared in a village Munak in Punjab (Now in Haryana) in the year 1855, 16 years after the irrigation system was operational. It was not an isolated case. In the year 1876, reports of waterlogging and soil salinity were made from Uttar Pradesh when an indigo farmer asked for compensation in lieu of the damages to his land on account of irrigation. Irrigation was introduced in the Nira canal command in Maharashtra during 1885. Reports of waterlogging and soil salinity appeared in the year 1888 and by 1905 an area of 1600 ha was taken out of cultivation (out of an irrigated area of 14,800 ha). The problem areas increased to 5570 ha in 1947. Twin problems of waterlogging and soil salinity were also observed within few years after the opening of lower Chenab canal in 1892.

Table 1.4 Irrigation potential (mha)

Plan	Potential			Utilization		
	Maj. & Med.	Minor	Total	Maj. & Med.	Minor	Total
At the end of						
Pre Plan upto 1951	9.7	12.9	22.6	9.7	12.9	22.6
First (1951-56)	12.2	14.0	26.2	11.0	14.0	25.0
Second(1956-61)	14.3	14.7	29.0	13.0	14.7	27.7
Third(1961-66)	16.6	17.0	33.6	15.2	17.0	32.2
Annual Plans(1966-69)	18.1	19.0	37.1	16.8	19.0	35.8
Fourth (1969-74)	20.7	23.4	44.1	18.7	23.4	42.1
Fifth (1974-78)	24.7	27.3	52.0	21.2	27.3	48.5
Annual Plans (1978-80)	26.6	30.0	56.6	22.6*	30.0	526
Sixth (1980-85)	27.7*	37.5	65.2	23.6*	35.2*	58.8
Seventh (1985-90)	29.9	46.6	76.5	25.5*	43.1	68.6
During 1990-91	0.4	1.8	2.2	0.5	1.7	2.2
During 1991-92(Ant)	0.6	1.9	2.5	0.6	1.7	2.3
At the end of 1991-92	30.9	50.3	81.2	26.6	46.5	73.1
Targets for 1992-93	0.7	2.0	2.7	0.7	1.9	2.6
Targets for Eighth Plan	5.1	10.7	15.8	4.3	9.4	13.7

– * Reappraised figures

Recent experience on this score also reveals an interlink between the introduction of irrigation and spread of waterlogging and soil salinity. Projects commissioned in the post independence period, also developed serious waterlogging and soil salinity problems. The waterlogging developed so fast in Tawa project in Madhya Pradesh that within few years of opening of the canal, farmers formed "*Mitti Bachao Abhiyan Samities*" (Save the Soil Committees). The water table in an area of about 100,000 ha is reported rising at an alarming rate in the newly constructed Indira Gandhi canal command. It is estimated that the areas under waterlogging and salinity in these projects are increasing annually. This calls for introduction of preventive technology at the time of planning of an irrigation project, close monitoring of the project and implementing drainage projects as soon as the problem shows first sign of appearance.

WATERLOGGING- DEFINITION AND CRITERION

In the simplest form waterlogging connotes restricted drainage, whether surface or subsurface. 'Clayton' has defined a land to be waterlogged when the water table is within 45-75 cm of the ground surface. The definition was first adopted in India, by the Drainage Board, in 1917. A modification to this definition was made in 1925 in the state of Punjab, when a land was designated as waterlogged, if it is thrown permanently out of cultivation due to rise in sub-soil water and is recorded as *banjar* in the revenue records. The watertable in the month of June should be within 1.25 m of the surface. A further modification in the definition was made to account for the yield of crop. It was decided to term a land to be waterlogged if the yield was reduced to 25 per cent of the optimum value, due to rise in sub-soil water. National Commission on Agriculture (1976) accepted the definition given by a Committee of Central Water and Power Commission. According to this definition 'a soil is waterlogged if the water table rises to an extent that root zone becomes saturated, diffusion of air is curtailed and amount of oxygen is reduced with increase in carbon dioxide'. Due to simplicity the criteria given by 'Clayton' is commonly used in India and abroad. However, present day drainage definition apart from extraction of excess water includes improvement of the water, air and temperature conditions in soils. It is now realized that differentiation be made between waterlogging due to water stagnation (land submergence) and waterlogging due to rise in the water table because strategies for the reclamation and management of the two groups of soil would differ. Different states follow different norms for defining waterlogging. Therefore, there may be some discrepancies in the estimates of the areas made by different agencies. Norms for some states are reported in Table 1.5.

Table 1.5 Norms for categorization of waterlogged areas in states

State	Water table depth below the ground level (m)
Uttar Pradesh (Sarda Sahayak Project)	
Worst Zone	Less than 1
Bad Zone	1–2
Alarming Zone	2–3
Safe Zone	More than 3
Punjab	
Very critical	0 – 1.5
Critical	0 – 2
Haryana	
Critical waterlogged	0 – 1.5
Critical	1.5 – 3
Karnataka	
Tunghbhadra Command	0 – 2
Himachal Pradesh	0 – 2
Maharashtra	
Fully waterlogged	Water at surface
Waterlogged	0-1.2

According to the Report of the Working Group (1991) set-up by the Ministry of Water Resources, following definition and norms for categorization have been accepted.

An area is said to be waterlogged when the water table rises to an extent that soil pores in the root zone of a crop become saturated, resulting in restriction of the normal circulation of the air, decline in the level of oxygen and increase in the level of carbon dioxide. The water table which is considered harmful would depend upon the type of crop, type of the soil and the quantity of water. The actual depth of water table, when it starts affecting the yield of the crops adversely, may vary over a wide range from zero for rice to about 1.5 m for other crops. The following norms are suggested :

Depth to water table (m)	Nomenclature suggested
< 2	Waterlogged
2 – 3	Potentially waterlogged
> 3	Safe

CAUSES OF WATERLOGGING

Waterlogging is caused both by natural and man made factors. Some of these are the following:

— Poor natural drainage as a consequence of unfavourable topography or unfavourable sub-soil geology like existence of hardpan at shallow depths.

— Heavy storm rainfall coupled with poor natural drainage.

— Spilling of rivers resulting in submergence of agricultural lands.

— Introduction of irrigation without taking into account characteristics of soils and sub-soils for their irrigability.

— Heavy losses of water due to seepage from canals, distributaries and watercourses.

— Poor on-farm water management resulting in poor application efficiencies.

— Development activities such as construction of roads, bridges, railway lines and buildings resulting in choking of natural drainage.

— Poor maintenance of existing drainage system and outlets.

— Deforestation and poor upkeep of watersheds.

— Hydraulic pressure of water from upper irrigated areas resulting in seepage outcrop in low lying areas.

IRRIGATION INDUCED WATERLOGGING

The waterlogging in irrigation commands could appear in the following forms.

— *Surface ponding* : Due to water stagnation on the land surface as a result of inadequate surface drainage

— *High water table* : Due to ground water levels rising and remaining in the root zone to adversely affect the crops because of inadequate ground water drainage.

Surface Ponding of Water

Surface ponding of water is a very common problem in monsoon climatic conditions and is experienced in almost all the irrigation commands. Analysis of rainfall data for various stations in the country reveals that such a possibility cannot be ruled out even in semi-arid or arid climatic conditions (Table 1.6). Degree of water stagnation and the periods for which water stagnates on the land surface, however, goes a sea change once irrigation system is introduced in an area. It is because capacity of the soil to absorb water reduces as a result of rise in water table. The developmental activities like the construction of canal net work, roads and railway lines in the irrigation project also obstruct the natural drainage as a result of which the degree and duration of water stagnation as well as recharge to the ground water are increased. Thus, it is necessary to plan for surface drainage improvements when irrigation projects are implemented.

Rising Water Table

India has a long history of environmental problems arising from irrigation. The first recorded incidence of waterlogging/ salinity is of 19th century in the area irrigated by the Western Yamuna Canal. By 1855 the salinity and alkalinity problems in this region became serious. Soil and water samples were taken to London for analysis in the Royal School of Mines. The analysis of these samples suggested a practical remedy that irrigation be accompanied by drainage. Incidence of alkali development was also reported from the area around Aligarh in the command of Ganga canal. Reports of salinity problems also came from the western state of Maharashtra where a reclamation programme was undertaken by Inglish and Gokhale (1927). In the state of Punjab soil surveys which were locally termed as *Thurgirdawari* were initiated. These surveys evaluated the extent of damage caused by salinity (*Thur*) or higher moisture content (*Sem*) in the root zone and were based on visual observation. Even in this century, so spectacular has been the rise of water table in some of the irrigated lands that used to be under dry farming only a few years back, have become unfit for cultivation of any upland crop during the rainy season. For example, parts of Rohtak district in Haryana where water table was more than 20 m deep experienced a rise of water table at the rate of 0.6 m per year reaching an average depth of 3.0 m in 1982. At Haryana Agricultural University farm, where the water table used to be at 15.6 m in 1967 at the time of introduction of irrigation under Bhakra Canal system had gone up to 2.0 m in 1984 (Kumar and Singh, 1994). In the Ukai-Kakrapar project whose first stage was completed in 1957-58 and second stage in 1972, the area where water table was within 1.5 m of the soil surface during the post-monsoon period went up to 29,971 ha in 1990 (Table 1.7). The area having water table within 3 m was 1,32,276 ha.

Table 1.6 Storm rainfall of various durations expected to occur at different probabilities

Station, climate and rainfall (mm)	Duration (days)	Storm rainfall (mm) at probabilities (percentage)				
		10	25	50	75	90
Hissar, arid	1	64	39	25	18	11
(375)	2	102	75	60	42	39
	3	119	93	75	60	52
	4	135	110	81	78	62
Ludhiana,	1	178	145	95	58	42
semi-arid	2	220	179	116	69	50
(681)	3	232	189	125	77	56
	4	237	195	132	85	66
Cuttack,	1	198	167	128	85	70
sub-humid	2	275	229	173	109	87
(1514)	3	293	246	188	123	101
	4	312	264	205	138	115
Dapoli, humid	1	363	309	227	166	140
(3372)	2	530	463	362	286	254
	3	623	555	452	376	342
	4	719	641	523	435	398

Table 1.7 Area under various hydro-isobaths (water table below soil surface) during post-monsoon season in Kakrapar project, Gujarat (ha)

Period	Depth of water table (m)				
	0-1.5	1.5-3.0	3.0-6.0	6.0-9.0	> 9.0
1957-58		1036	49663	195148	160152
1989-90	29971	102305	147997	55425	*

* Area included in 6.0-9.0 m range
Note: The area during 1989-90 does not include *Kharlands* and non-project command (NPC) which were included in pre- irrigation survey

Even in the driest part of the country where irrigation was introduced in 1961 through Indira Gandhi Canal Project (IGNP) in Rajasthan, a steep rise in water table has been reported varying from 0.6 m/annum in Ganga canal to 1.0 m/annum in IGNP (Hooja et al., 1994). The introduction of irrigation through lift irrigation schemes was also affected with this malady although only limited quantities of water were introduced. In the command area of Jui Canal constructed to supply water to the arid parts of Haryana, there has been continuous rise of water table (Sharma and Bhargava, 1982) ranging from 0.3 to 0.8 m per year. The compilation of data by Singh (1993) confirms that once canal irrigation is introduced, the water table rise is inevitable (Table 1.8).

EXTENT OF WATERLOGGING

The precise estimates of the extent of waterlogging are not available. There is no universally accepted criteria to characterize a land as waterlogged. Attempts have been made to estimate the waterlogged area by different agencies at different points of time. The estimates of the National Commission on Agriculture (1976) appear to be the most appropriate. Out of nearly 6 mha area affected due to waterlogging, 3.4 mha was due to surface flooding and 2.6 mha was due to rise in water table. The last figure closely matches with the recent estimate of the Working Group constituted to identify waterlogged areas due to rise in the water table in different irrigation commands (Table 1.9). The data reported by National Commission on Agriculture (1976) covered surface stagnation of water and rise in water table both in commands and the humid regions of the country.

Table 1.8 Rate of rise in water table in different irrigation commands

Irrigation command	Rate of rise in water table (m/year)
Mahi Right Bank Canal Command (MRBC), Gujarat	0.28
Rajasthan Canal Command (IGNP), Rajasthan	0.29 - 0.88
Western Yamuna & Bhakra Canal Command, Haryana	0.30 - 1.00
Sirhand Canal Command, Punjab	0.10 - 1.00
Sarda Sahayak Canal Command, Uttar Pradesh	0.68
Malprabha Canal Command, Karnataka	0.60 - 1.2
Nagarujan Sagar Irrigation Project, Andhra Pradesh	0.32
Sriram Sagar Irrigation Project, Andhra Pradesh	0.26

IRRIGATION INDUCED SOIL SALINITY

Soil salinity problems in irrigation commands develop whenever soil and hydrological conditions favour the accumulation of soluble salts in the root zone. The rise of water table in semi-arid and arid areas mobilises the salts present in the soil profile and ground water. Once the ground water table rises between 2 and 3 m of the soil surface, it contributes substantially to evaporation from the soil surface and water uptake by plants. The upward flux of water due to evaporation and water uptake by plants, results in gradual concentration of salts in the root zone. In the initial years, the crop yield may be reduced due to salinity, but as the severity of the problem increases with time, the lands may have to be abandoned.

According to recent estimates, the salt affected soils occupy about 8.5 mha of lands mostly in arid and semi-arid regions besides an appreciable area in the humid coastal tract (Singh, 1992, FAI, 1994). Distribution of salt affected soils in the country is shown in Fig. 1.1. It is to be noted that the area reported in this table is for those soils which have already

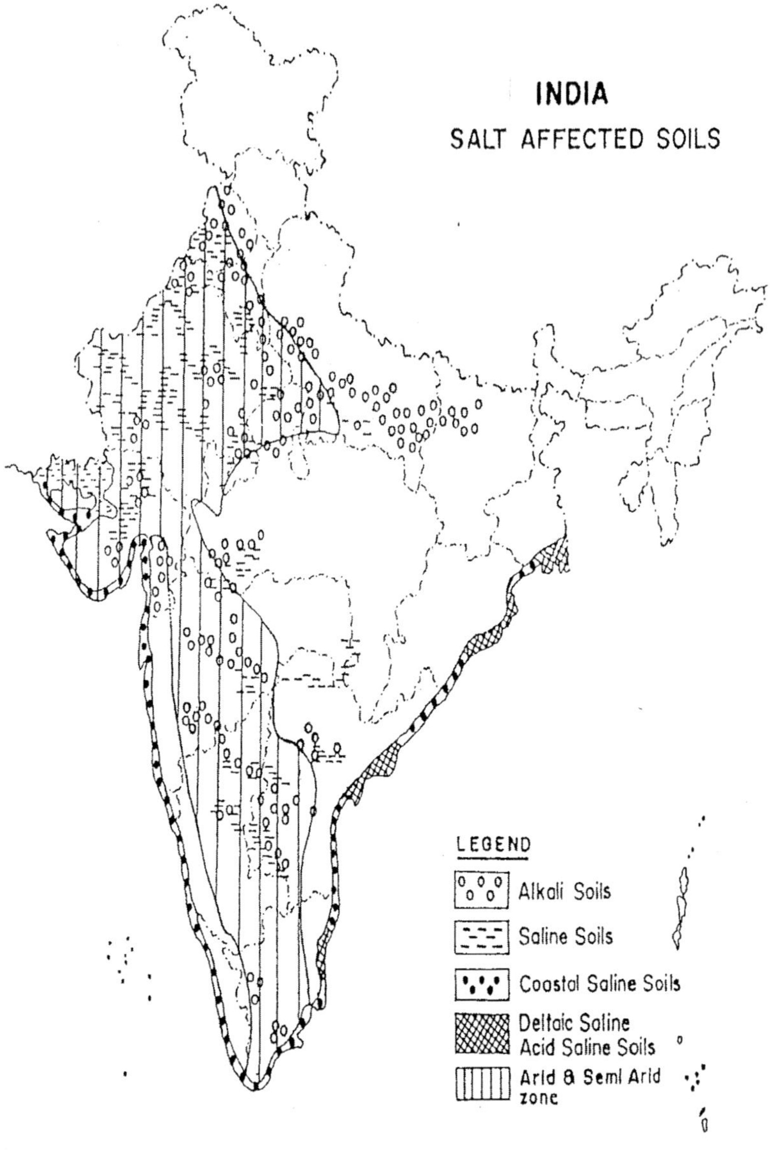

Figure 1.1 Distribution of salt affected soils in India

gone out of cultivation. There could be large areas where salinity problem has tended to affect crop yields.

Table 1.9 Extent of waterlogged area (lakh ha)

State	National Commission on Agriculture, 1976	Reconciled estimate of the Working Group, 1991
Andhra Pradesh	3.39	2.66
Assam	N.R.	N.R.
Bihar	1.17	6.20
Gujarat	4.84	1.72
Haryana	6.20	2.49
Jammu & Kashmir	0.10	0.01
Karnataka	0.10	0.24
Kerala	0.61	0.12
Madhya Pradesh	0.57	0.73
Maharashtra	1.11	0.15
Orissa	0.60	1.96
Punjab	10.90	2.00
Rajasthan	3.48	1.80
Tamil Nadu	0.18	0.16
Uttar Pradesh	8.10	4.30
West Bengal	18.50	N.R.
Delhi	0.01	N.R.
Total (lakh ha)	59.86	24.56
(mha)	6.00	2.46

Source: Report, Working Group (1991). N.R. = Not reported.

CAUSES OF SOIL SALINITY

Salt affected soils may develop from normal soils through the accumulation of salts in one of the following ways. A combination of these processes could be operative at several places.

- Accumulation of salts from applied irrigation water
- Movement of salts from lower layers
- Upward movement of ground water through capillary rise
- Sea water inundation in coastal regions
- Deposition of wind blown salts

TYPES OF SALINITY PROBLEM

Conventionally, according to USSL (Richards *et al.* 1954) the salt affected soils have been grouped as alkali soils, saline soils and saline-alkali soils (Table 1.10). However, considering their specific management requirements these soils have been identified in two main classes namely (i) alkali, and (ii) saline soils (Fig. 1.2 and Fig. 1.3).

Alkali Soils: These soils contain excess of salts like bicarbonates, carbonates and silicate of sodium, capable of alkaline hydrolysis or have sufficient exchangeable sodium to interfere with growth of most crops. They have pH_s greater than 8.5, electrolyte conductivity of saturation extract less than 4 dS/m and exchangeable sodium percentage (ESP) of 15 or more. It is estimated that nearly 3.45 million ha of land resource is afflicted with the problem of alkalinity covering alluvial and black soils mainly in Punjab, Haryana, Uttar Pradesh and Madhya Pradesh (Table 1.11). Physico-Chemical characteristics of a alkali pedon are reported in Table 1.12.

Fig. 1.2 A view of alkali land in the Indo-Gangetic plain

Fig. 1.3 A view of the inland waterlogged saline soil in Mahi-Kadana command in Gujarat

Table 1.10 Classification of salt affected soils on the basis of chemical characteristics (USSL, 1954)

Calss	EC_e (dS/m)	ESP	pH	Indian Local Names
Saline Soils	> 4	< 15	< 8.5	*Thur, Uippu, Lona, Shora, Soula, Pokhali, Khar and Kari*
Saline-alkali soils	> 4	> 15	Variable	*Usar, Kallar, Karl, Chopan, Bari, Reh, Choudu and Kshar*
Non-saline alkali	< 4	> 15	8.5-10.0	*Usar, Rakkar, Bara, Chopan Kari*

Table 1.11 Distribution and characteristics of alkali soils in India

Region	Nature of problem	Area affected (000 ha)	Soil Characteristics	Ground water quality
Indo-Gangetic	Alkali	2169	High pH, medium EC, presence of $CaCO_3$ in sub soil, free carbonates in soil solution, restri -cting layers in sub-soil.	Usually good quality, occasionally high RSC waters
Medium and deep black soils	Alkali	1275	Neutral to alkali pH, medium to high salt content with chlorides and sulphates, presence of smectites, high CEC, calcareous	Variable often high EC, Mg:Ca may be more than 1.
Total		3444		

Table 1.12 Physico-chemical characteristics of an alkali soil from Etah (U.P.) representing Indo-Gangetic plain

Depth (cm)	EC_e (dS/m)	pH_s	Ionic composition (meq/L)						
			Ca+Mg	Na	K	CO_3	HCO3	Cl	SO_4
0-16	13.84	10.2	2.2	210.0	0.2	151.0	28.0	14.0	21.0
16-27	12.50	10.1	2.5	190.0	0.2	129.0	20.0	14.0	13.0
27-64	4.60	9.7	1.3	45.0	0.1	32.0	10.0	11.0	3.0
64-81	1.92	9.8	2.5	19.0	0.1	4.0	13.0	5.0	1.0
81-108	1.88	9.4	2.0	16.0	0.1	4.0	9.0	6.0	1.0
108-127	2.35	9.1	1.0	21.0	0.1	7.0	10.0	6.0	1.0
127-148	1.40	8.6	2.7	13.0	0.1	2.0	6.0	6.0	1.0

Table 1.12 contd....

Depth	Sand	Silt	Clay	Textural class *	Ca CO_3 %	OM** %	CEC cmol (p^+) kg^{-1}	Soil ESP
	%							
0-16	62.0	24.0	14.0	sl	1.9	0.7	6.7	88
16-27	57.0	25.5	17.5	sl	4.0	0.6	7.3	87
27-64	45.0	22.0	33.0	scl	3.8	0.6	15.9	67
64-81	38.0	35.0	27.0	l	2.7	0.6	13.9	78
81-108	34.0	34.0	32.0	cl	12.0	0.5	12.9	24
108-127	39.5	33.0	27.5	c	23.8	0.5	12.0	31
127-148	65.0	18.0	17.0	sl	25.0	0.4	8.0	22

* sl sandy loam, scl sandy clay loam, l loam, cl clay loam, c clay
** OM organic matter

Saline Soils: These soils contain neutral salts (chlorides and sulphates of sodium, calcium and magnesium) in quantities enough to interfere with growth of most crop plants. The pH of saturated soil paste is less than 8.5, soil EC_e more than 4 dS/m and ESP less than 15. Problem of soil salinity is widespread in the arid and semi-arid region (irrigated lands) and sub-humid to humid regions (coastal salinity). A total of 5.01 million ha area is estimated to be affected by soil salinity (Table 1.13). Inland salinity is widespread in alluvial, black, red and grey soils.

From practical point of view, there are following three distinct groups of saline soils :

i) Inland saline soils with shallow water table which is generally saline. These soils suffer with waterlogging and high salinity.

ii) Normal soils being irrigated over the years with saline and/or sodic underground water. These soils normally have deep water table. However, in certain situations the water table could be shallow. These soils generally occur in arid and semi arid regions.

iii) Coastal saline soils where the problem is due to ingress of sea water resulting in high salinity and temporary flooding. The climate of these areas varies from humid to semi-arid.

Physico-chemical characteristics of a inland saline soil are shown in Table 1.14. The differences in the physico-chemical characteristics are quite apparent (Table 1.12 and 1.14).

Table 1.13 Distribution and characteristics of saline soils in India

Region	Nature of problem	Area affec ted (000ha)	Soil characteristics	Ground water quality
Medium and deep black	Saline soils	972		
–do–	Saline	887*	Neutral to alkali medium to high salt content, mostly chlorides and sulphates of Na, free carbonates absent.	Usually poor, high EC, occasionally high RSC
Red and grey	Saline	1097	Neutral to slightly alkaline pH, high salt content of chlorides due to sulphates of Na, Mg: Ca often exceeds 1.	Usually saline or alkaline waters with presence of toxic elements
Coastal saline soils	Saline	2104	Neutral to slightly alkaline pH, medium to high salt content, prone to monsoon flooding.	Shallow saline, free water scarce in summer
Humid tropics	Acid sulphate	35	Acid pH, presence of humic pyritic horizon, high EC_e.	Shallow saline
Total affected area		5095		

* Partly water-logged,** Includes coastal districts
Source: Singh (1992)

NATURE OF SALINE SOILS

These contain neutral salts (chlorides and sulphates of sodium, calcium and magnesium) in quantities enough to interfere with growth of most crop plants.

Inland saline soils are extensive in arid and semi-arid regions. Mean annual rainfall in this region is less than

Table 1.14 Physico-chemical characteristics of a saline soil from Jind district (Haryana)

Depth (cm)	pHs	EC_e (dS/m)	Ionic composition (meq/L)							Boron (mg/L)
			Ca+Mg	Na	K	CO_3	HCO_3	Cl	SO_4	
0-11	7.5	4.5	20.5	35.6	0.6	Nil	1.0	35.5	21.5	11
11-30	7.6	4.6	30.7	22.6	0.5	Nil	1.0	40.5	10.0	6
30-56	7.8	4.8	32.6	23.6	0.3	Nil	1.0	37.5	12.7	6
56-76	7.8	5.6	30.5	27.3	0.2	Nil	1.0	39.5	14.5	7
76-141	8.0	8.3	41.5	42.8	0.2	Nil	6.0	58.0	22.5	10
141-200	8.0	12.7	50.7	74.1	0.2	Nil	10.0	70.5	39.5	14

Depth (cm)	Sand	Silt	Clay	$CaCO_3$ (%)		CEC cmol (p^+) kg^{-1}	OM (%)	ESP
	%			< 2 mm	> 2 mm			
0-11	69.3	21.2	9.5	Nil	–	4.4	0.5	6.8
11-30	61.5	24.9	13.6	0.8	–	5.7	0.3	7.0
30-56	60.9	21.8	17.3	2.9	–	7.8	0.3	7.6
56-76	55.7	21.8	22.5	4.4	7.0	8.9	0.2	8.9
76-141	57.0	21.6	21.4	8.0	35.0	8.9	0.2	8.9
141-200	56.1	22.0	21.9	16.8	15.0	7.6	0.1	9.2

550mm. Mostly these soils occupy low to intermediate topographic situation. These have a zone of salt accumulation at shallow depths due to limited leaching under prevailing climatic conditions. If the existing natural equilibrium is disturbed by introduction of canal irrigation, the resulting water table rise sets in the process of salinisation. Depending on genetic soil heterogeneity like soil texture, structure, presence of diagnostic horizons and kind of salts present, saline soils have some discrepancies among themselves by way of calcic, non-calcic, petrocalcic, and gypsic soil types. Salient features of a typical calcic saline soil are as follows:

These are very gently sloping soils developed on quaternary alluvium. Soil matrix colours vary from yellowish brown to brown. Yellowish red coloured mottles are noticed in the upper part of the profile indicating release of ferruginous oxides during mineral weathering and subsequent redistribution. Soil texture varies from sandy loam in the surface to sandy clay loam or clay loam between 50 and 110 cm depth, while a platy structure in surface few cm changes to angular or subangular blocky before turning massive at 1m or deeper depths. Discrete calcium carbonate nodules occur at a depth from 60 to 150cm forming a calcic horizon. Iron-manganese concretions and yellowish brown mottles commonly occurring above the calcic horizon indicate periodic cycles of excessive wetting followed by soil drying.

Illitic mineralogy and saline environment imparts stability to soil aggregates which do not disintegrate. Therefore, in general the soil exhibits better infiltration and percolation. Fine texture and higher bulk density in subsurface horizons between 60 and 100cm depth reduces percolation rates and soils exhibit extremely low hydraulic conductivity.

NEED FOR RECLAMATION OF SALINE WASTELANDS

Demand for agricultural products is increasing on account of increasing population and improving standard of living. With liberalization of economy demand scenario has further changed as a result of additional responsibilities for generation of increasing surpluses for conversion to value added products for domestic use and exports. Natural resources in two major areas that is land and water are rapidly shrinking because of keen competition between various sectors of economy and as a result of degradation. Increasing pressure on these resources is accelerating the degradation processes.

Developing strategies that will promote conservation of resources, which will reclaim the degraded resources and ensure the much needed productivity gains would be crucial in our future efforts to increase agricultural production. Problem specific technologies, therefore, would play an increasingly important role in the coming decades. The book deals with basic principles of reclamation and management of waterlogged saline wastelands and describe a number of technologies for water table control and reclamation and management of saline waste lands.

REFERENCES

Fertilizer Association of India (FAI) 1985. **Soils of India and their Management**. Fertilizer Association of India, New Delhi. 445 p.

Hooja, R., Srinivas, V. and Govind Sharma. 1994. Waterlogging and possible remedial measures in the Indira Gandhi canal command area development project. **Natn. Seminar on Reclamation and Management of Waterlogged Saline Soils.** CSSRI, Karnal.

Khosla, A.N. 1949. Appraisal of water resources- Analysis and utilization of data. **Proc. U.N. Scientific Conf. on Preservation and Utilization of Resources**. UNESO, New York.

Kumar, R. and Singh, J. 1994. Drainage system for ground water management. **Natn. Seminar on Reclamation and Management of Waterlogged Saline Soils**. CSSRI, Karnal.

Ministry of Water Resources, 1990. **National Water Policy.** Central Water Commission, New Delhi, p. 11.

Murthy, R.S. et al. 1982. **Benchmark Soils of India**. National Bureau of Soil Survey and Land Use Planning, Nagpur. Indian Council of Agril. Res., New Delhi. 374 p.

National Commission on Agriculture (NCA), 1976. Report of the Commission, Climate and Agriculture (Part IV). **Ministry of Agriculture and Irrigation.** Govt.of India, New Delhi.

Inglish, G.C. and Gokhale, V.K. 1928. Report on land drainage experiment at Baramati, **Technical paper No. 24,** Bombay Public Works Department.

Irrigation Commission, 1903. A Report. **Ministry of Irrigation**, Govt. of India, New Delhi.

Raghava Rao, K.V., Raju, T.S. and Ramesan,V. 1969. An estimate of ground water potential of India- first approximation. **Soil Water Management Symposium,** Hisar. 6-16.

Rayachaudhari, S.P. 1962. **Land Resources of India**. Planning Commission. New Delhi.

Report of the Working Group on Problem Identification in Irrigated Areas with Suggested Remedial Measures, 1991. **Ministry of Water Resources,** Government of India, New Delhi.

Richards, L.A. (Ed.) 1954. Diagnosis and Improvement of Saline Alkali Soils. **USDA Handbook 60**, 160 p.

Sharma, S.K. and Bhargava, G.P. 1982. **Annual report**. CSSRI, Karnal.

Singh, N.T. 1992. Salt affected soils in India. In **Land and Soils** (Khoshoo, T.N. and Deekshatulu, B.L. Eds.). Indian National Science Academy, New Delhi, 65-100.

Singh O.P. 1993. Drainage problems and design criteria for land drainage symptoms. **Proc. National Workshop "Sustainable Irrigation in Saline Environment"** CSSRI, Karnal 204-218.

2

LABORATORY AND FIELD INVESTIGATIONS

The success of reclamation and management of salt-affected soils depends very much on the correct diagnosis of the nature and intensity of the problem. Accordingly, representative soil and water samples must be adequately characterized in the laboratory and in the field. Collection of representative soil and water samples is, therefore, basis for the correct diagnosis and for making sound recommendation for management of problem soils. Commonly employed field and laboratory investigations for the diagnosis of the problem have been discussed.

SOIL SAMPLING PROCEDURE

How to Sample Soils

The method for soil sampling depends on the purpose for which a sample is collected. For this reason, no common standard procedure can be suggested for all the situations. To meet the requirements of a good sample, some useful points with particular reference to soils anticipated to have a salinity or alkali problem are discussed below.

Number of Samples

The size of plot or the farm area that could be represented by one composite sample, depends on the expected spatial variability in the fields. Field plots which are more or less uniform may make a single unit for the purpose of sampling. Normally 8 to 10 samples are collected from a uniform field unit to make one composite sample representative of the area. However, salt affected soils have lot of spatial variability. Therefore, when it is intended to bring large acreage of barren lands under reclamation, it is appropriate to level and bund the field and then take surface (0-15cm) and subsurface (15-30cm) sample for laboratory analysis. Thus, for each ha field 2-3 composite samples should be sufficient. For this purpose a thin 1/2 to 3/4 slice of soil taken from 20 sites spread over 0.5 ha field, should be thoroughly mixed in a tray to draw a composite sample for laboratory analysis. When it is intended to improve a patchy field, model soil samples should be composited from a few representative samples taken from within the spot.

Depth of Sampling

The depth to which samples should be obtained for analysis depends on the land use. If the intention is to grow shallow to medium rooted crops like rice, wheat etc. it is sufficient to obtain samples upto a depth of 30 cm. A major fraction of the roots of these crops is confined to this zone. If, on the other hand, it is proposed to grow deep rooted crops like cotton, it is advantageous to obtain samples upto a depth of about 60 cm. In case the intended land use is to grow trees (horticultural or forest tree species) it is essential to carry out a detailed study of the soil profile upto a depth of 150 to 180 cm. This is necessary because the performance of tree species on a long term basis is strongly influenced by properties of the lower soil layers including the presence of a hard layer due to $CaCO_3$ concretions etc.

How to Obtain Samples

For samples upto 30 cm depth, a cut in the soil can be made with a spade and a thin slice of soil taken to a desired depth with the help of trowel (Mason's spatula) (Fig. 2.1). If sample from deeper soil layers are needed, an auger should be used (Fig. 2.2). Alternatively, a soil profile can be exposed to a depth of 1.5 m, and sample can be taken from appropriate

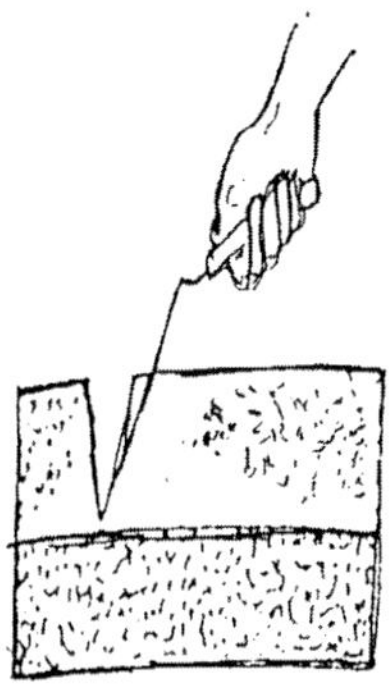

(a) A thin slice of soil should be taken with the help of a trowel or Khurpi

(b) Thoroughly mix on a dry cloth or in a tray

(c) Reduce the sample weight by neglecting sample marked x and if necessary repeat

(d) Bag and send for analysis

Figure 2.1 Soil sampling procedure using a trowel

depths. Once the profile is exposed, consult a specialist for examination and formulation of recommendations.

How Much Sample

For conventional diagnostic analysis in a laboratory, it is sufficient to have about 500 to 750 g of soil sample.

Figure 2.2. A view of soil sampling technique with an auger

Important

The slick spots can be best identified during the cropping season. It is better to sample when spots are visible. Salt crusts should not be mixed with the soil samples. Collect salt crust separately. Samples representing slick spots should not be mixed with samples from good spots. Ensure that the samples are not taken immediately after rains/irrigation. Also avoid sampling near a irrigation channel, drainage channel, bunds, farm yard manure pits, spots under a shady tree and from threshing floors or roads.

Packing of Samples

Before packing of the samples and their mailing to a soil testing laboratory, following precautions may be taken.

- Remove stones/gravel from samples.
- Air dry the samples in trays in shade.
- Air dried soil should be packed in a polythene/hard paper bag.
- Insert the polythene bag in the cloth bag and tag an appropriate label.
- Cloth bags should also be labelled with copying pencil.
- Seal the bags properly.
- Use wooden/hard paper board cartons for mailing.

Each sample should be accompanied by a detailed proforma to help make sound recommendations on the basis of soil analysis (Table 2.1). A sample sheet on which soil analysis results are provided to the user is given in Table 2.2 (Gupta *et al.* 1995).

Table 2.1 Collector's observations on the soil sample

Specifications	
Name of the farmer	
Name of the collector	
Location of the site	
Any other character	
Proposed cropping pattern	
Observations (if already in use)	
On crops (give yield in kg/há) Amount of water applied General observations	
Type of soil	
Soil fertility test required	Yes/No*
Whether water sample is being sent	Yes/No*
Sample No. and date	
Name of the laboratory where sample sent Signature, address and date	

* Encircle the appropriate word

Table 2.2 General proforma for soil test reports

Name of the laboratory	______________
Name of the farmer	______________
Sample No.	______________
Date of Sample received	______________
Sample analyzed	______________

Description	Results				
ECe (dS/m)	[ ]				
pH	[ ]				
Cations	Ca^{++}	Mg^{++}	Na^{+}	K^{+}	Total
(meq/L)	[ ]	[ ]	[ ]	[ ]	[ ]
Aminos	CO_3^{--}	HCO_3^{-}	SO_4^{--}	Cl^{-}	Total
(meq/L)	[ ]	[ ]	[ ]	[ ]	[ ]
Special determination	ESP	GR (t/ha)	Nitrogen	Phosphorus	Potash
	[ ]	[ ]	[ ]	[ ]	[ ]

Analyzed by

Checked by

Recommendation by (if any)

Signature

Water Sampling

Proper diagnosis of the nature and the amount of salts present in a water is essential to assess the possible hazards which it could pose to the soil or the crop. Collection of representative sample is, therefore, basic for correct diagnosis and for making sound recommendations for the management of poor quality waters. Following procedures may be adopted in collection and despatch of the samples for analysis.

Selection of Containers

As far as possible plastic bottles should be used for collection of water samples. Glass bottles can also be used if plastic bottles are not available. Bottles should be thoroughly cleaned and rinsed 2-3 times with water before taking the samples. avoid using metal containers for collection of water samples.

Time of Sampling

In case of gravity canal, take the sample of the water from the channel after irrigation has been in progress for about an hour. For ground waters, take the sample only after running the tubewell for about 0.5 hr.

Quantity of Sample

Most routine salinity appraisal tests require about 100 ml of water. However, to be on the safer side and to allow specialized tests, if need be, a minimum of 500 ml of sample should be sufficient.

Packing

Water containers should be properly screwed and if possible wax sealed. Containers should be properly labelled for identification. Each sample should be accompanied by a description conforming to the list (Table 2.3).

Table 2.3 Collector's observations on the irrigation water sample

Specifications	
Name of the farmer	
Name of the collector	
Location of the site	
Source of water	Spring/canal/drain/ well/lake*
Additional details (if needed)	
Colour	
Odour	
Any other character	
Depth of water on sampling date	
Proposed application	Domestic/irrigation*
Irrigation method employed	surface/sprinkler/drip*
Proposed cropping pattern	
Observations (if already in use)	
On crops (give yield in kg/ha)	
On soil	
Amount of water applied	
General observation	
Type of soil	
Your opinion on water quality	
Any special test you require	
Whether soil sample is being sent	Yes/No*
Sample No. and date	
Name of the laboratory where sample sent	
Signature and date	

* Encircle the appropriate word

WHERE TO SEND SAMPLES FOR ANALYSIS ?

In each state, the Department of Agriculture/the State Agricultural University has soil/water testing laboratories either at the district or at the *tehsil* level. These laboratories are generally responsible for undertaking soil and water testing and making appropriate recommendations, either free or

Table 2.4 General proforma for water test reports

Name of the laboratory ______________________

Name of the farmer ______________________

Sample No. ______________________

Date of Sample received ______________________

Sample analyzed ______________________

Description	Results					
ECe (dS/m)	[ ]					
pH	[ ]					
Cations (meq/L)	Ca^{++} [ ]	Mg^{++} [ ]	Na^{+} [ ]	K^{+} [ ]	Total [ ]	
Aminos (meq/L)	CO_3^{--} [ ]	HCO_3^{-} [ ]	SO_4^{--} [ ]	Cl^{-} [ ]	Total [ ]	
Special determination	Boron (mg/L) [ ]	Fluorine (mg/L) [ ]	Silica (mg/L) [ ]	NO_3 (me/L) [ ]		Any other [ ]

Analyzed by ______________________ Checked by _____

Recommendation by (if any) ______________________

Signature ______

on a nominal charge. The sample is analyzed by a trained chemist. Each department/ University/ laboratory has its own set of proforma for reporting the analytical results along with their recommendations. A specimen proforma is reproduced in Table 2.4.

MEASUREMENT OF DEPTH TO WATER TABLE

The main purpose of any investigation on ground water is to get information on the position, seasonal and annual fluctuations of the ground water which will determine the severity and extent of the drainage problem. Many a times vegetative cover is a good index to depth to water table in an area. Certain trees and grasses are more common in areas with high water table. It may be a useful practice to look for these features while surveying the area. Trees such as willow, cottonwood and poplar and grasses such as reed grass and hedges often thrive in high water table areas. Scientific investigations for quantitative estimates are carried out using ground water table data from open wells, observation wells and piezometers.

Water Table in Open Wells

Open well is a very common feature of rural India. These wells have been an important source of water to meet the domestic and agricultural water requirement. The water table in these wells at a particular time of the year and in different seasons will reveal, the ground water table behaviour in the region. Central Ground Water Board/ State Directorate of Ground Water have fully utilized the potential of these wells in monitoring ground water table behaviour. Usually pre-monsoon and post-monsoon ground water table is measured by these departments. Recently, few ideally located wells have also been selected to undertake monthly reading and for water quality analysis. These data provide wealth of information to a designer. It is common experience that the existing wells have concentration at few points while in other the density of wells is quite sparse (Fig 2.3). Thus, there is need to install more observation wells/ piezometers to minimise gaps in information and to have a better insight into the regional problem. A rough guide to decide the number of wells/observation wells is given in Table 2.5.

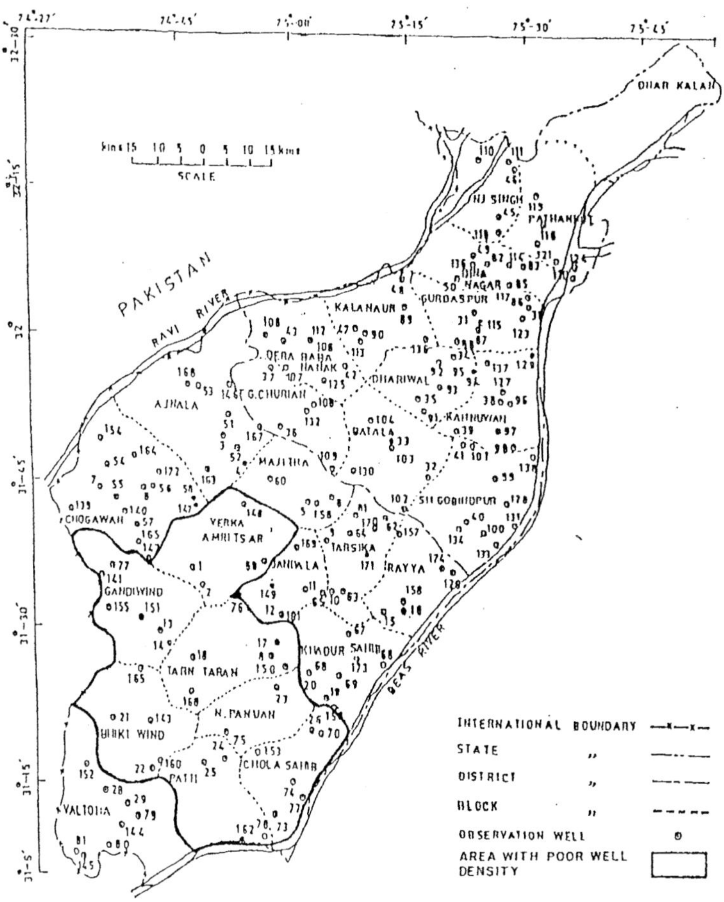

Figure 2.3 A view of the distribution of wells in a *Doab* in Punjab.

Table 2.5 Wells/Observation wells required in relation to the size of the drainage area

Size (ha)	No. of wells/ observation wells	No./100 ha
100	20	20
1000	40	4
10000	100	1
100000	300	3

Installation of Observation Wells/Piezometers

An observation well consists of a pipe that is perforated from bottom to about 60 cm below the ground surface (Fig. 2.4a). The water level in an observation well represents the average hydraulic head over the length of the perforated portion of the pipe. At shallow depths, this head corresponds with the watertable surface. The simplest form of an observation well (though of temporary nature) is an auger hole left uncased in the field which may be employed to measure watertable depth during an exploratory survey.

The perforations in the observation well are drilled in five to six rows and are covered with jute or coir rope, or muslin cloth to prevent fine particles from entering the pipe or clogging the tiny holes. The lower open end of the pipe is sealed with a plug. The pipe is placed in the centre of a bore hole made with a hand or power auger and the annular space between the pipe and wall of the hole over the entire length of the perforated portion is filled with gravel pack or coarse sand (Fig. 2.4a). This gravel pack facilitates the flow of water from the soil into the pipe and vice versa. The rest of the annular space is filled with fine materials (clay, clayey fine sand) whose permeability is low enough to prevent leakage along the pipe. To keep the pipe in position and to prevent any direct contact between surface water and the backfill material, a concrete seal is provided around the pipe, extending from the ground surface to a depth of about 15 cm. The top of the pipe is fitted with a cap to prevent entry of soil into the pipe.

Piezometers are used to measure the hydrostatic pressure at different depths in the sub- soil layers. Though both observation wells and piezometers are required in the initial stages, piezometers are installed selectively. Initially, a battery of piezometers is installed at one location with the lower end of each piezometer terminating at different depths. The number is increased if the problem of artisan pressure is indicated. Piezometers are reliable tools for determining hydraulic head at a particular depth while the observation wells are used to measure fluctuations of the watertable levels. A piezometer is an open-ended nonperforated galvanized iron (or sometimes plastic) pipe of 4 to 5 cm diameter, driven or jetted into the ground to the depth at which the hydraulic head is to be measured. The water level in the pipe corresponds with the hydraulic head at the lower end where the pipe terminates. To

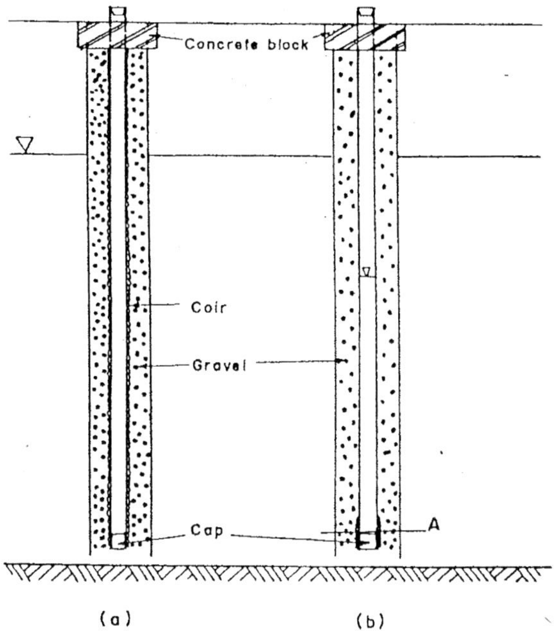

Figure 2.4 Schematic cross-section through (a) a observation well and (b) a piezometer

ensure that the piezometer functions properly, the lower 0.5 m of the pipe may be perforated (Fig. 2.4b).

LABORATORY INVESTIGATIONS FOR QUICK DIAGNOSIS

Electrolyte Conductivity (EC)

The total salt content of the soil solution or of the irrigation water is represented in 3 ways i.e. milligrams per liter (mg/L), milli equivalents per liter (meq/L) or as electrolyte conductivity expressed as dS/m. Although EC has the units of Siemens/m, it is commonly reported in deci-Siemens/m (dS/m) which is equivalent to the earlier unit expressed as millimhos/cm (mmhos/cm) (1 dS/m = 1 mmhos/cm). Milliequivalent per liter is the milliequivalent weight of the anion or compound (1/1000 part of the equivalent weight) in a liter of solution. Electrolyte conductivity is by far the simplest representation and is a measure of the ease with which current will flow through the soil water or the irrigation water. All the 3 units are related with each other. The relation between meq/L and EC is

$$meq/L = 10 \times EC \qquad 0.1 < EC < 5.0 \qquad (2.1)$$

Relation between mg/L and EC is

$$mg/L = 0.64 \times EC \qquad 0.1 < EC < 5.0 \qquad (2.2)$$

At higher EC (greater than 5 dS/m), the value of the factor in eq. 2.2 increases and tends to be near unity at very high salinity levels. These relations are only approximate and have been obtained by plotting the values for several salts commonly occurring in the soils. Ionic composition in meq/L is converted to mg/L by multiplying with its equivalent weight.

Electrolyte conductivity has the advantage of simplicity since it can be readily measured in the field or laboratory by using a conductivity meter. It is closely related to the sum of cations (or anions) as determined chemically and usually correlates closely with the total dissolved solids. As the electrolyte conductivity would depend upon the moisture content and temperature; it is common to express the conductivity at common moisture and temperature. The total soluble salt concentration is estimated by measuring the EC of the saturation extract at 25°C.

Conductivity meters are used to measure electrolyte conductivity in the laboratory. These meters are usually provided with an immersion type conductivity cell with platinum electrodes. A 0.01 N potassium chloride solution (0.7456 g of potassium chloride dissolved in distilled water and made to 1 L at 25°C) is used as a standard solution to calibrate the electrode. This solution at 25°C has an electrolyte conductivity of 1.412 dS/m. Most conductivity meters are provided with temperature correction. If such correction is not provided, electrolyte conductivity so determined should be corrected for temperature and multiplied with cell constant to be evaluated using the standard solution as shown below.

$$\text{Cell constant} = \frac{\text{EC of the standard solution at } 25^{\circ}\text{C}}{\text{Measured EC of the standard solution at } 25^{\circ}\text{C}}$$

In many soil testing laboratories for quick appraisal of soils, EC is usually determined in 1:2 soil water extract (EC_2) which is prepared by dissolving 10 g of dry soil with 20 g of water. EC_2 is highly and significantly correlated with EC_e (Gupta et al. 1969, Singhla et al., 1974). A calibration curve relating EC_2 and EC_e is used to estimate EC_e. A typical curve

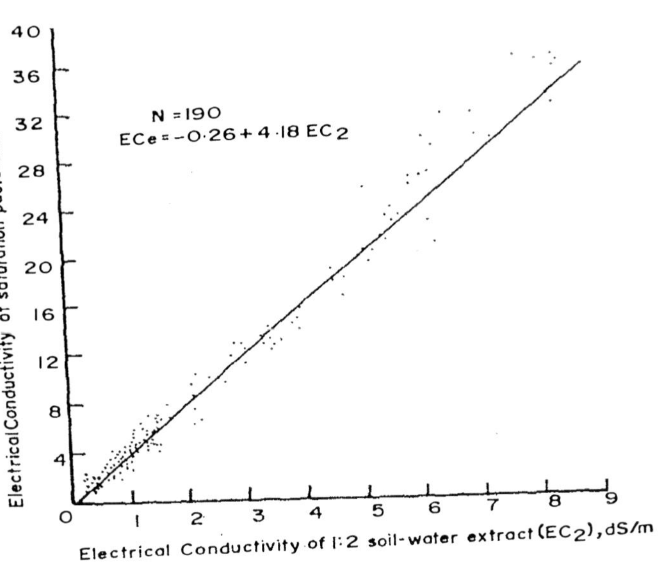

Figure 2.5 Relationship between EC_2 and EC_e for a medium textured soil

used for this purpose for a medium textured soil is given in Fig. 2.5 (Sharma, 1994). It shows that for the soil used in calibration, EC_e is nearly 4 times the EC_2.

Soil Reaction, pH

The pH scale is a measure of the effective concentration of the hydrogen ion. pH is the log of the inverse of the hydrogen ion concentration in the saturated paste of the soil (pH_s). It is used as an index of the acidity or the alkalinity of the soil. A pH of 7.0 is neutral. Soils are acidic below pH 7.0 and alkaline above pH 7.0. A pH_s of 8.5 or more precisely 8.2 or above characterises soil as alkali (sodic). Soil water suspensions (1:2) are used to determine pH, when rapid determinations are needed.

A pH meter with glass electrode is used to measure the pH of the saturated soil paste (pH_s) or pH of the 1:2 soil water extract (pH_2). Soil pH_2 and pH_s are significantly correlated (Gupta, 1972). It is desirable to use the same procedure regularly so that the values are consistent and have maximum diagnostic value. It is observed that pH of the alkali soils of the Indo-gangetic alluvial plains is well correlated with the ESP values. Therefore, pH has been used as a diagnostic tool for appraisal of alkali problem and for evaluating gypsum requirement.

Exchangeable Sodium Percentage (ESP)

Each soil has certain capacity to adsorb cations (the positively charged constituents of the dissolved salts) such as calcium (Ca^{++}), magnesium(Mg^{++}), sodium(Na^{+}) and potassium (K^{+}). It is known as cation exchange capacity (CEC) and is expressed as c mol (p^{+}) kg^{-1} of soil. Depending upon the concentration of these ions in the soil solution, they get adsorbed. Once the relative concentration of these ions changes in the soil solution, the cations could be replaced by each other although the capacity of the soil remains the same. Exchangeable sodium percentage is the amount of adsorbed sodium, expressed in per cent of the CEC of the soil

$$ESP = \frac{\text{Sodium on the exchange complex (c mol } (p^{+}) (Kg^{-1}))}{\text{CEC (c mol } (p^{+}) \text{ kg}^{-1} \text{ soil)}} \times 100 \quad (2.3)$$

The regression equation (eq. 2.3) is valid in cases where EC of the water is less than 5 dS/m and Ca/Mg ratio more than unity.

Sodium Adsorption Ratio, SAR

Ratio of the monovalent cations and the divalent cations is used as an index of the exchangeable sodium or ESP. On this basis, an index called as sodium adsorption ratio is given as below. As all lons are expressed as meq/L, units of SAR are $(meq/L)^{1/2}$.

$$SAR = \frac{(Na)}{\sqrt{(Ca + Mg)}} \quad (2.4)$$

SAR of the irrigation water is commonly used to calculate the SAR of the soil solution(SAR_e) and ESP of the soil. For steady state relation between SAR_W and SAR_e

$$SAR_e = SAR_W/\sqrt{LF} = SAR_W \; CF = SAR_W \sqrt{\frac{ECe}{EC_W}} \quad (2.5)$$

Where LF is the leaching fraction i.e. fraction of the water applied drained below the rootzone. Gupta (1987) reported that SAR_e was highly correlated with SAR_W (Table 2.6). ESP and SAR_W are related as follows (U.S. Salinity Laboratory Staff, 1954)

$$ESP = \frac{100\,(-0.0126 + 0.01475\,SAR_W)}{1 + (0.0126 + 0.9475\,SAR_W)} \qquad (2.6)$$

Table 2.6 Observed and calculated values of SAR of the irrigated soils

Site	Irrigation water		Irrigated soil		
	EC(dSm^{-1})	SAR	EC(dSm^{-1})	Obs. SAR	Calc. SAR
Dantiwada	5.7	18	10.9	26	25
Kaparda	9.8	25	11.8	31	28
Jelwa	5.3	21	6.7	21	23
Shikarpura	2.3	9	22.2	32	28
Surajata	8.7	7	11.1	6	8
Dhakri	7.5	20	14.4	28	28

The relationship between ESP of soils and SAR of the soils/waters holds good only for such soils which have exchangeable sodium ratio (ESR = ES/(CEC–ES)) value less than 1.0, corresponding to ESP of less than 50.

The regression equation as listed above is limited to waters of EC less than 5 dS/m and Ca/Mg ratio greater than unity.

Sodium to Calcium Activity Ratio, SCAR

In situations where EC is higher than 5.0 dSm^{-1} and Mg/Ca ratio is higher than 1.0, SAR_W is very poorly correlated with ESP of the soils. However sodium to calcium Activity Ratio (SCAR = $Na/\sqrt{Ca}$) was observed to be highly correlated with ESP of the soils (Gupta, 1987) (Table 2.7).

Table 2.7 Correlation between soil ESP, SCAR and SAR of the soil irrigated with saline water

Location	Irrigation Water			Irrigated soil		
	EC (dSm^{-1})	Mg/Ca ratio	SAR	SCAR	SAR	ESP
Kaparda	10.8	4.4	28	59	31	60
Jelwa	5.9	8.1	23	36	28	38
Shikarpura	4.5	16.0	8	26	32	35

DIAGNOSIS OF SALINE AND ALKALI SOILS

While survey of an area is in progress, some visual observations would reveal the nature of the problem. The visual characters that could act as a diagnostic criteria in identifying the two groups of soil during a reconnaissance survey are reported in Table 2.8. Exact analysis should, however, be based on the chemical characteristics as revealed by the sample analysis. The differentiating parameters are described in Table 2.9.

Table 2.8 The visual observations for distinguishing between alkali soil and saline soil

Observation	Alkali soil	Saline soil
Rain water stagnation	Stagnation for longer period	Relatively Lesser period
Stagnated water type	Dark coloured, muddy soapy	Clear
Surface salt crust	Dark brown or ash coloured clay crust	White salt efflorescence
Ground water	Usually good, occasionally with high Residual Sodium Carbonate (RSC)	Saline, sometimes with high Sodium Adsorption Ratio (SAR)
Presence of nodules	$CaCO_3$ nodules are present at some depth	Gypsum may be present.
Natural vegetation (grasses)	*Sporobolus marginatus, Desmostachya bipinata, Suaeda maritma, Diplachne fusca, Dichanthium annulatum, Cynodon dactylon, Chloris barbata Brachiaria mutica, Kochia indica, Panicum antidotale*	*Cressa cretica, Cyperus rotundus, Chloris pallida, Sporobolus pallidus, Haloxylon salicornicum, Aclurопus lagopoides, Zygophyllum simplex, Dichanthinm annulatum, Sueda fruticosa, Butea monosperma*

Table 2.9 Characteristics of alkali and saline soils

Characteristics	Alkali soils	Saline soils
EC_e (dS/m)	Variable	4.0 or more
ESP/SAR	15 or more	Variable
pH_s	8.2 or more	< 8.2
Chemistry of soil solution	Carbonates and bicarbonates always present	Dominated by chloride and sulphate
Main adverse effect on plants	Alkalinity of soil solution	High osmotic pressure of soil solution
Geographical distribution	Associated with semi-arid and sub-humid climates	Associated with arid and semi-arid climates
First step for reclamation	Lowering or neutralization of high pH using amendents	Removal of excess electrolytes through leaching

WATERLOGGED SALINE SOILS

Although the basic framework for reclamation and management of saline and waterlogged saline soils would be similar, waterlogged saline soils would require water table control measures before actual reclamation could start (See Chapter 3).

REFERENCES

Gupta, I.C. 1972. Note on the relationship between pH of saturated soil paste and 1:2 soil water suspension. **Indian J. agric. Sci.** 42: 962 - 963.

Gupta, I.C. 1986. Quality of irrigation water-recent criteria and classification. **Curr. Agric.** 10: 1-42.

Gupta, I.C. 1987. Predicting sodification in saline water irrigated Soils. **J. Indian Soc. Soil. Sci.** 35 : 169-170.

Gupta, I.C., Bohra, R.K. and Abichandani, C.T. 1969. Relationships between certain indices of saline water irrigated soils of western Rajasthan. **Ann. Arid Zone** 9 : 228 - 230.

Gupta, S.K., Dinkar, V.S. and Tyagi, N.K. 1995. **Reclamation and Management of Waterlogged and Salt Affected Areas in Irrigation Commands-A Manual.** Ministry of Water Resources, New Delhi and CSSRI, Karnal. 144 p.

Sharma, D.P. 1994. Personal Communication.

Singhla, S.K., Bhargava, G.P. and Gupta, I.C. 1974. Relationships between certain diagnostic indices of salt affected soils of the Indo-Gangetic alluvial region. **Ann. Arid Zone** 13 : 218-223.

U.S. Salinity Laboratory Staff. 1954. **Diagnosis and Improvement of Saline and Alkali Soils.** U.S. Deptt. Agric. Handbook 60. 160p.

3

ROOT ZONE SOIL SALINITY

Secondary soil salinisation in agriculturally productive lands has become a serious issue as nearly 18 % of the net irrigated area through major and medium irrigation projects has gone out of cultivation because of excess soluble salts. Other issues apart, timely recognition of salt accumulation in the root zone is becoming increasingly important. High irrigation efficiencies are sought to face the impending shrinking water resources as well as to minimize drainage water disposal related problems. Thus, due to reduced leaching fraction, risk of salinity build-up is much more than before. Moreover, with competitive demand for fresh water from other sectors of economy, poor quality of waters would increasingly find application in agriculture requiring regular monitoring of the root zone for maintaining its productivity. The present chapter describes salt balance components that need monitoring to diagnose the problem. Processes that are operative during salinity build-up under different set-ups are discussed.

SALT BALANCE OF THE ROOT ZONE

The water and salt balance of the root zone are so intricately woven that it is not possible to separate them out. The salt storage in the root zone of irrigated lands can be worked out by various components of water shown in Fig. 3.1 by multiplying with their respective salt concentrations (Gupta and Khosla, 1996). The resulting salt storage or leaching from the root zone can be mathematically put into the form of an equation as follows:

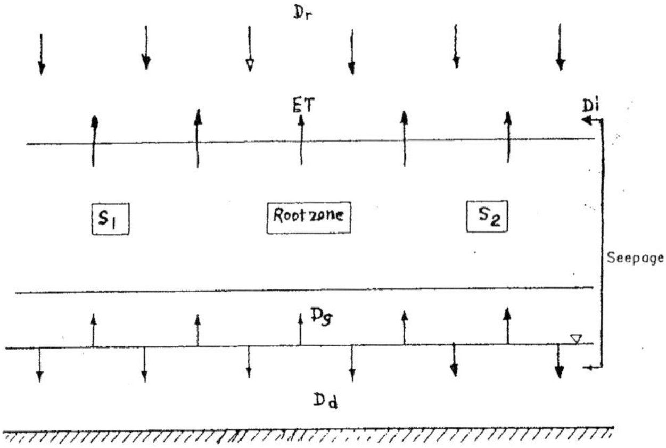

Figure 3.1 Water balance components of the root zone of irrigated lands

$$S = D_rC_r + D_iC_i + D_gC_g + S_1 - D_dC_d - S_2 \qquad (3.1)$$

In eq. (3.1) salt storage, S, in the root zone is given by components which add salt in the root zone (+) and those which remove the salt (-). These components are added together with their respective signs. On the plus side are the salts added through rainfall (D_rC_r), the application of irrigation water (D_iC_i), the contribution of ground water (D_gC_g) and added directly to the root zone such as through application of amendments/fertilizer or generated in the root zone such as from mineral dissolution (S_1). On the minus side are the salts removed as a result of drainage (D_dC_d) and precipitation (S_2). In most cases, the two terms given by S_1 and S_2 are neglected leaving only four terms on the right hand side of eq. (3.1). It may be remembered that irrigation induces mineral dissolution and in many cases, it may not be possible to neglect this component without causing wide differences in the calculated and actual salt balance. It is clear from eq.(3.1) that to control salinity build-up in the root zone either one has to induce drainage from rainfall and applied irrigation water and/or to minimize the salt addition to the root zone. As rain water is usually a good quality water, the salt added could only be controlled either by reducing the salts applied with irrigation water or through the control of ground water which might add substantial amounts of salts in the root zone in case ground water is at shallow depths. This task is easier said than done. For example, application of irrigation water or the contribution of the ground water to meet the crop water requirement can not be controlled at will as it may adversely affect the crop as a result of inadequate supply of water. Moreover, unknown parameters in eq. (3.1) add another dimension to the complexity in the use of this equation. For example D_g, and C_d are not exactly known and D_d can only be estimated. Therefore, it is always useful to monitor the salt status of irrigated lands over a few years time to work out the salt regime. It would help not only in the diagnosis of the problem but also in identifying factors which are likely to accentuate salt accumulation or those which would help to leach down the salts.

FACTORS AFFECTING ROOT ZONE SALT BALANCE

Soil salinity in the root zone is dynamic and has large spatial and temporal variability. Some of the factors that influence the salinity of the root zone are the amount of water applied, the salinity of the irrigation water, depth to water table, soil texture and other soil characteristic which affect its permeability, evapotranspiration, total rainfall and its distribution, irrigation methods, setting of the land in respect of the adjoining area and above all the overall management inputs that go with any agricultural activity. The following salt balance situations may occur:

i. Balance in favour of leaching (S in eq.(3.1) is negative)

ii. Stable salt balance (S in eq. (3.1) is zero)

iii. Balance in favour of salt accumulation (S in eq. (3.1) is positive)

The salt balance is determined for short durations (say a month), medium durations (say a season) and long term durations (say annual) over a few years time. Similarly, salt balance is established over small terrestrial units (say a farm) as well as on regional basis. Salt transport from irrigated lands and its accumulation in unintended areas is schematically shown in Fig. 3.2. Such processes operating on regional scale generally justify investments on assessing regional salt balance along with smaller units. While preparing the salt balance sheet, causes, rate and degree of accumulation or leaching should also be worked out.

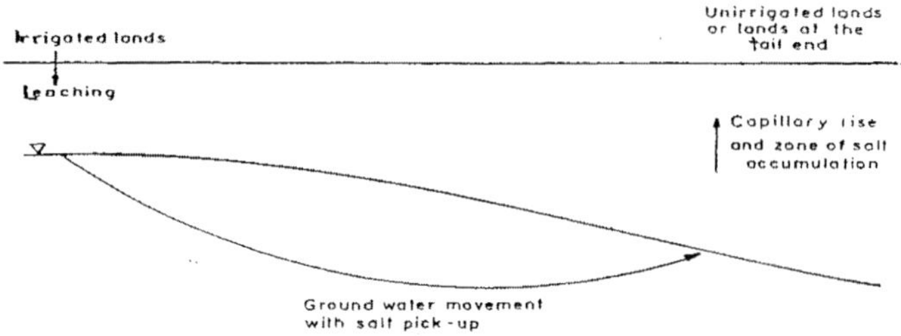

Fig. 3.2 Salt balance scenario in irrigated lands showing areas with salt leaching and salt accumulation

CAUSES OF SOIL SALINITY

Salt affected soils may develop from normal soils through the accumulation of salts in one of the following ways. A combination of these processes could be operative at several places.

- Accumulation of salts from applied irrigation water
- Movement of salts from lower layers
- Upward movement of ground water through capillary rise
- Sea water inundation in coastal regions
- Deposition of wind blown salts

These processes are briefly described in the following paragraphs.

Irrigation with Saline Waters

Soils which are otherwise salt free may be rendered unproductive by use of irrigation water containing excess of salts. In irrigated lands particularly when limited quantities of water are applied, the salts continue to accumulate in irrigated soils if salts are not removed in equivalent amounts as are applied with irrigation water. The general characteristics of salinisation-desalnisation cycle in a saline water irrigated land are shown in Fig. 3.3.

(a) The period of soil salinity build up (Zone A) during November to April with each irrigation till equilibrium is reached.

(b) Period of salt redistribution (Zone B) within the root-zone from April to mid June or till the onset of monsoon. During this period, most of the times, soil salinity is maximum in the upper surface layer (0-15 cm).

(c) Period of desalinisation (Zone C) during monsoon season when in most places rainfall is in excess over the potential evapotranspiration. This results in net leaching of salts.

Five factors control the extent to which salinity may develop in the soil profile. These are : (i) quality of irrigation water, (ii) number of irrigations, (iii) soil texture (iv) leaching fraction and (v) rainfall.

Quality of Irrigation Water

Soil salinity (EC_e, the electrolyte conductivity of the saturation extract) and ground water quality used for irriga-

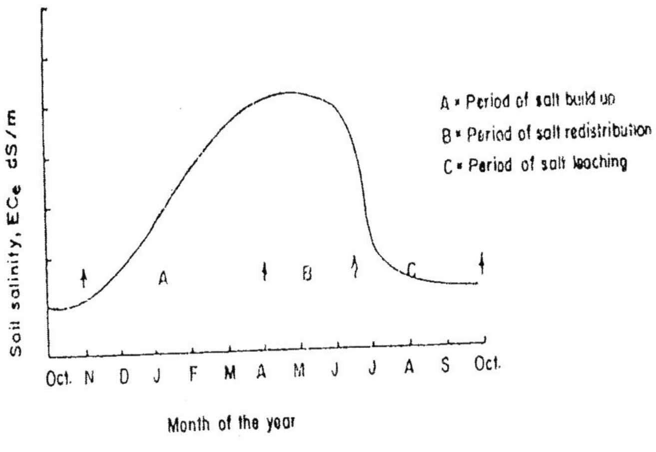

Fig. 3.3 Salinisation desalinisation cycle in a saline water irrigated land

tion (EC_i) are related to each other. A relationship which relates the two could be expressed in the form of a regression equation as given below:

$$EC_e = a + b\ EC_i \qquad (3.2)$$

Where, 'a' and 'b' are regression coefficients. The regression coefficients of eq. (3.2) for four stations show large variations (Table 3.1). This variation could be due to variations in the management conditions or as a result of the variations in other four factors affecting soil salinisation- desalinisation cycle.

Number of Irrigations

The regression constants of the regression eq. (3.2) for the Agra station are presented in Table 3.2 after 2,3,4 and 5 number of irrigations. The data show that the constant 'b' which really matters in such situations, varies from a low value of 0.66 for two irrigations to a value of 1.21 for 5 irrigations. The average value for 3-5 irrigations is 1.02 while for 2-3 irrigations it is 0.83. It is observed that there would be little change in soil salinity after 5-6 irrigations, if other condition remain unaltered. A regression equation which takes into account the irrigation water quality as well as number of irrigations is of the following form:

$$EC_e = -2.26 + 0.90\ EC_i + 1.24\ N. \qquad (3.3)$$

Here, N is number of irrigations (Gupta, 1986). Equation (3.3) can be used in areas with medium texture soil and where number of irrigations is not more than 6.

Soil Texture

Build-up of soil salinity is more in heavy than light textured soils. The regression coefficient, b, varies from 0.36 for sandy to loamy sand group of soils but is 0.69 for fine loam, clay loam and silty clay loam group of soil under similar situations. The value of 'a', may not vary much for the two groups (2 to 2.2).

Table 3.1 Regression constants of eq. (3.2) for four stations

Station	Regression constants of eq. (3.2)	
	a	b
Agra[1]	1.59	1.02
Dharwar[1]	0.50*	0.24*
Jobner[2]	3.87	0.48
Pali[2]	0.84	0.44

[1]Data from experimental stations. [2] Data from farmers' fields
(*) Regression constants for EC_2 versus EC_i (EC of the irrigation water)

Table 3.2 Regression coefficient as influenced by number of irrigations

Number of Irrigations	Crop	Regression constants of eq.(3.2)	
		a	b
2	Wheat	1.54	0.66
3	Wheat	2.14	0.90
4	Wheat	2.07	1.00
5	Wheat	0.62	1.21
3-5	Wheat	1.59	1.02
2-3	Mustard	2.00	0.83

Leaching Fraction

Leaching fraction which is defined as the fraction of the irrigation water that leaves the root zone can be easily known from the average water application efficiency, if distribution efficiency is reasonably high. An application efficiency of 70 per cent means that leaching fraction is around 0.30. The values of K of the following relationship are : 2.0 for low, 1.0 for medium and 0.5 for a high leaching fraction (The qualitative terms, low, medium and high leaching fraction are for 0.1, 0.3 and 0.5, respectively). It also means high (90 per cent), medium (70 per cent) and low (50 percent), water application efficiencies. The values of K could be still higher for still lower leaching fractions. This can also explain the effect of soil texture on soil salinity. As for the same amount of water applied, the leaching fractions will be more in light textured than a heavy textured soil.

$$EC_e = K\ EC_i \tag{3.4}$$

Amount of irrigation water applied and the soil texture effects could be clubbed together in a single factor that is leaching fraction. Depending upon the leaching fraction, mean root zone salinity of a soil could be as low as half of the salinity of the irrigation water to as high as 4 times the salinity of the irrigation water. Thus, there is ample scope to control the mean root zone salinity at a pre-decided level within a wide range by controlling leaching fraction. For the same mean root zone salinity, crop production at or near to optimal levels could be possible, if the upper portion of the root zone is somehow kept relatively salt free as the crops adjust for their water requirement and draw more water from non-saline zone. However, over time when salinity build-up proceeds upward, the yield of crops would be affected.

Rainfall

Rainfall has a two fold effect on soil salinity build up. High rainfall not only reduces the number of irrigations, but being a very good quality water also leaches the salts accumulated in the root zone. The effect of rainfall on the regression constants of eq. (3.2) for the Agra station shows that with an additional 10 cm of rainfall, the constant 'b' is reduced by about 0.6 unit for both the crops i.e. wheat and mustard (Table 3.3).

Table 3.3 Effect of rainfall on regression constants of eq. (3.2) for wheat and mustard crop

Crop	Rainfall (mm)	Number of irrigations	Regression constants	
			a	b
Wheat	17.1	4	2.44	1.29
Wheat	113.9	3	1.76	0.69
Mustard	18.2	3	2.76	1.11
Mustard	113.9	3	1.80	0.50

Movement of Salts from Lower Layers

Mineral soils are derived largely from the weathering of rocks. There is a close relation between salt accumulation and the chemical composition of rocks from which soils are formed. Soils formed directly from salt-bearing rocks usually contain excessive salts. In some cases these salts under natural condition lie at some depth. When aided by man's activities such as introduction of irrigation, these salts are transported to the root zone.

Ground Water and Capillary Rise

Ground water in arid and semi-arid regions often lies at deeper depth. At such depths, there is no fear of water transport due to capillary rise. The water table is in equilibrium with natural recharge and drainability of the land. This equilibrium is disturbed with the introduction of irrigation resulting in more recharge to the aquifer. As a consequence, the water table starts rising such that in a few years time it could be within the capillary fringe. Due to capillary rise, water evaporates at the land surface, leaving behind salts in the root zone. Large areas have been adversely affected by soil salinity/ alkali in irrigation commands owing to this process. The total quantity of water brought in and the concentration of the ground water determine the extent of soil salinisation.

Critical Depth to Water Table

The concept of 'Critical Depth to Water Table' has been in vogue for quite sometime since it was first advocated by Talsma (1963). According to this concept, secondary salinisation would occur as long as the water table is above the critical depth. The critical depth is defined as the depth to water table at which evaporation at the soil surface through capillary rise reduces to 1/10 of its rate at which water would evaporate in case water table is at or near to the soil surface. For practical purposes, the rate should not exceed the maximum allowable flux. The allowable maximum flux to maintain favourable water and salt balance may vary from 0.5 mm/day to 1 mm/day. On this basis, critical depth may vary from a meter in light textured soils to a few meters in fine textured soils.

Recent studies have revealed the limitations of this concept and arguments contesting the applicability of this concept have been put forth. It is argued that if the flux at the bottom of the root zone could be kept downward, there should not be any fear of soil salinisation resulting from the shallow water table. In the Indian context, shallow drains have been found to work well in maintaining water and salt balance in drained lands because flux of the water at critical periods is downward (monsoon season,when the water table is at shallow depths) and during the summer season when evaporative demand is high, water table is deep enough to allow excessive upward flux. As a result, applicability of the concept of critical depth to water table needs to be examined before it is applied to field situations.

Soil Texture and Upward Flux

Water is the main carrier of salts in the soil. As the soil solution reaches the site of evaporation at or near the soil surface, the water is evaporated and the salts are left behind. In cropped soils, the zone of salt build-up depends upon depth and distribution of active roots and the fraction of water that reaches the soil surface. Upward water flux decreases with increase in depth to water table. The generalized relationship between water table depth and upward water flux suggests that water can certainly move at some appreciable rate from water tables at least as deep as 5 m. It is observed that the evaporation rate decreases more steeply with water table depth in a coarse than a fine textured soil. The maximal limiting upward flux (cm/day) is given by the following relationship:

$$q_{max} = \frac{A\ a}{d^n} \qquad (3.5)$$

where d is the depth to water table (cm), a (cm^{n+1}/day) and n are coefficients and A is also a constant which depends on n. Some typical values of these constants are presented in Table 3.4 as also the maximum upward flux.

Ground Water Quality and Permissible Maximum Flux

For maintaining favourable water and salt balance, understanding and knowledge of the maximum flux values is most important. For good quality ground water a value of 0.5 mm to 1 mm/day has been suggested and conventionally 1 mm/day is taken as the limiting value. In general, as the ground water quality deteriorates, value of q_{max} is reduced to minimize soil salinisation.

Salinity in the profile increases with decreasing depth to ground water table but increases with increasing salinity of the ground water at the same depth to water table. The effect of ground water salinity (0.6-3.7 dS/m) is more pronounced at shallower than at deeper depths. The studies have shown that co-relation between the concentration of the salts at the soil surface and the ground water is significant only when water table is at or above 1.2 m.

Table 3.4 Experimental values of constants of eq. (1.4)

Soil texture	n	(cm^{n+1}/day)	A	Max. upward flux* (cm/day)
Sand	4	1.7×10^8	1.52	2.58
Fine sandy loam	3	3.2×10^5	1.76	0.56
Loam	2	1.7×10^3	2.46	0.42
Clay	2	1.1×10^3	2.46	0.27

*For d=100 cm

Capillary Rise and Soil Salinity Build-up

In soils with a shallow water table, yield is depressed due to reduced soil aeration and inhibited root extension. Salt build-up may further reduce the yields. For the same amount of water applied through irrigation or drawn from the shallow ground water to meet the crop requirement, salinity build-up would be more with water drawn from the ground water as there would be no leaching in this case. The distribution of salts in the root zone in the two cases would also be entirely different. As a result of salt distribution pattern, crops would suffer more in the latter than in the former case even if the mean root zone salinity is the same. Thus, in cases where techniques which aid the crop to draw upon water from the shallow water table are employed, no doubt, it would reduce the number of irrigations, lower the production costs and decrease the drainage volumes and attendant problems, but at the same time it would increase the chances of soil salinisation. Thus, under this situation provision for leaching to reduce salts below the permissible limit should be made in advance.

Inundation with Sea Water

In coastal areas, land inundation with sea water is often a factor leading to soil salinisation. Sea water may enter through the creeks during high tides. In these areas shallow ground water of poor quality could be another factor contributing to soil salinity.

Wind Blown Salts

Another way of soil salinity build up particularly in coastal belts could be the deposition of salts by winds. Huge quantity of salt is transported by winds when travelling over the sea or over the salt heaps which gets deposited over the adjoining lands. Such a process of soil salinisation has been very common in Kutchh region of Gujarat.

TIME FRAME

In irrigated agriculture, time frame within which salt balance of the root zone is determined could be very important not only from the point of view of saving water but also for the crop health. The time frame would normally depend upon how fast the salinity build-up occurs in the root zone. For salt sensitive crops irrigated with saline water, salt balance at the end of each irrigation may be important. For salt tolerant crops, it may be possible to allow build-up of the salt in the root zone and carry out leaching at appropriate time when water is available. In the monsoon climatic conditions where annual salinisation-desalinisation cycles occur, zero or negative salt storage should be aimed at on a yearly basis. Within a year, there could be periods of salt build-up depending upon the tolerance of the crop.

SUSTAINABILITY OF IRRIGATED AGRICULTURE

As mentioned elsewhere in this chapter, serious doubts have been raised on the sustainability of irrigated agriculture in recent times. Past experiences where civilizations have been ruined are quoted to emphasize the ill effects of irrigated agriculture. Without doubt, past experiences where irrigated

agriculture not only failed to bring prosperity but ruined the potential beneficiaries could be cited to bring home the point yet there is ample scope to ponder over as to what had gone wrong and where. One could learn from these instances and rectify the mistakes that have been made. Today, we have much more information based on field experiences and fully supported by theoretical exercises assuring us that there is no serious threat to permanence of irrigated agriculture. However, this threat could be real, if man fails to use the practices and knowledge now available to him to devise location specific ways and means to fight the menace of waterlogging and soil salinity. To assess long term sustainability of irrigated agriculture, one could also resort to modelling for scenario building exercises to arrive at interventions that would minimize salinity build-up. Any failure of irrigated agriculture at this stage would be the failure of the man to put into practice present day knowledge available on the subject. In any case, it would not mean doom of irrigated agriculture.

REFERENCES

Gupta, S.K. 1986. Salinisation-desalinisation in saline water irrigated soils. **Proc. Seminar on Problems of Waterlogging and Salinity in Alluvial Soils.** New Delhi. 12.01-12.16.

Gupta, S.K. and Khosla, B.K. 1996. Salinity control in the root zone of irrigated soils. **Proc. Workshop Waterlogging and Soil Salinity in Irrigated Agriculture. CBIP, New Delhi and CSSRI, Karnal.** 163-180.

Talsma, T. 1963. **The Control of Saline Groundwater.** Meded Landb Hoogesch, Wageningen. 63 1-68.

4

WATER TABLE CONTROL FOR RECLAMATION

From management point of view, the waterlogged lands can be classified into three groups as shown below :

- Waterlogged lands
  - Surface water stagnation
  - High water table
    - Good ground water quality
    - Poor ground water quality

Saline lands are often associated with waterlogged areas due to rise in the water table linked with saline/sodic ground water. During the monsoon season, surface stagnation of water could also be a serious problem in some areas occupied by saline soils as rain water can not be absorbed in adequate quantities due to shallow water table. Thus, both surface and subsurface drainage are needed to minimize adverse effects of waterlogging. Control of the water table in these lands is necessary and is achieved through the installation of some kind of subsurface drainage. Horizontal drainage appears to be the most promising for this purpose. Improvement in surface drainage, on the other hand would protect the crops from adverse effects of water stagnation as well as help to minimize the cost of subsurface drainage. Herein, appropriate surface and subsurface drainage technologies that relate to the reclamation of waterlogged saline lands, are discussed.

Drainage Nomenclature

On the basis of site from where the water is to be removed, drainage has been classified into two groups:

i. Surface drainage

ii. Subsurface drainage

In the first case, removal of water is sought from the surface of the land. On the other hand in subsurface drainage, removal or control of ground water is the prime objective. In many cases control of salinity may also be sought using water as the transporting agent. Most common sub-classifications of the subsurface drainage, however, have been based on material/structure used for draining purposes. On this basis subsurface drainage is classified as below.

a. Pipe drainage system

b. Tubewell drainage system

c. Bio-drainage system

As the names suggest, in the first case cement concrete tiles, plastic pipes or pipes of any other suitable materials are used for drainage while in the second case tubewells are sought to be used for draining the lands. The name bio-drainage has been derived from the fact that a biological living material, a exotic tree, is used as the drain material. Other names for the three classes are the horizontal drainage system and the vertical drainage system. While horizontal drainage system is used to designate the pipe drainage system; tubewell and bio-drainage, respectively, are designated by the name vertical drainage. The names are derived as the drain-axis in the first case travels along the soil surface (horizontal) while in the second and third case it is perpendicular to the soil surface (vertical). Some other classification may also be prevalent but we restrict here to the nomenclature discussed above.

SURFACE DRAINAGE

Problem of Surface Drainage

Flat or nearly flat areas, heavy soils with low permeability and lands in humid or tropical and sub-tropical regions where high intensity storms are common, are subject to surface inundation and therefore, require surface drainage. Surface drainage problem many a times, may not be of local origin as in low lands it may be due to the ingress of surface runoff from uplands. Overflow from rivers or natural channels sometimes contributes largely to the drainage problem of an area. The following factors contribute to the problem:

i. Uneven land surface causing hinderance in the natural outflow of run-off from the area.

ii. Inadequate capacity of the drainage channels during critical periods.

iii. Inadequate outlet conditions.

iv. Back water flow.

Surface drainage is carried out to meet one or more of the following objectives:

i. In flat areas for timely removal of water so as to allow optimum growth of plants,

ii. In sloping areas for orderly removal of water such that it does not cause erosion and

iii. For the collection and disposal of surface irrigation run-off.

Surface drainage is often considered as a panacea to all drainage problems. However, it is known that it can not help to lower the water table effectively although build-up of the water table could be minimized through efficient surface drainage. During initial years of reclamation, although rain water conservation helps leaching down the salts, but in the long run surface drainage would minimize subsurface drainage flows. Thus, operational cost of drainage could be minimized.

Parts of the system

A surface drainage system essentially consists of the following three parts:

i. Collection system

ii. Disposal system

iii. Outlet

Collection system mainly concerns with on-farm drainage with disposal and outlets forming a part of the overall system. A major component of the disposal system and the outlets is taken care of by the State Governments as these are designed, executed and maintained by the concerned department.

Drainage Coefficient

Drainage coefficient is defined as the amount of water that runs off from a given area and is to be removed within 24 hr. Commonly the unit is depth of water per day but many a times, the drainage coefficient is also expressed as flow rate per unit area. It is often not possible to determine drainage coefficient with any degree of accuracy although it is the key factor in establishing the needed capacity of the drainage system.

Several methods are used to evaluate the drainage coefficient. The use of a particular method mainly depends on the availability of required data. The most simple method of assessing surface drainage coefficient for farm drainage consists of the following steps (Finkel, 1983) :

i. Evaluate maximum 24 hr rainfall depth which may occur with a probability of 10-20 per cent. Frequency analysis procedures are commonly employed for this purpose. A computer programme to carry out this analysis is given in Annexure I. It uses normal, lognormal, SMEMAX and Gumbel distributions. Readers may choose one of the distributions for use depending upon the correlation coefficient which indicates the degree of the fit of the distribution to the observed data. These coefficients are also received as an output of the computer programme.

ii. Evaluate the basic infiltration rate of the soil and estimate the potential evapotranspiration.

iii. Estimate the tolerance of crop to surface inundation. For example let us say it is 'n' days. Multiply the infiltration rate and expected potential evapotranspiration by the number of days, 'n'.

iv. Subtract the value calculated in step (iii) from the value in step (i). The resulting value when divided by number of days, n will give the drainage coefficient in depth of water/day.

These steps can be described mathematically in the form of an equation which is given below:

$$q = \frac{1}{n} [R - n(E + I)] \qquad (4.1)$$

Here q is the drainage rate in mm/day, R is the rainfall in mm, E and I are potential evapotranspiration and infiltration rate in mm/day and n is number of days. When n is greater than one, it may be useful to increase the duration of rainfall in step (i) also from 1-day maximum to n-days maximum. With this change, drainage coefficient will be more closer to reality. Similarly, for the rice crop, as some depth of water is allowed to stand in the fields after the drainage period, necessary correction in eq. (4.1) could be made by subtracting this depth while calculating the drainage coefficient for this crop. A number of workers have estimated drainage coefficient for different agro-climatic regions of the country. Most comprehensive analysis made by Gupta *et al.* (1971) revealed that drainage coefficient in India varies from a low value of 2 mm/day to about 40 mm/day depending upon the crop (rice, grain or vegetable) and the climatic conditions.

SURFACE DRAINAGE REQUIREMENT AND CROPS

Not all crops suffer equally from inundation or root submergence. The damage which is dependent upon the degree and duration of submergence varies from one crop to another and even varieties within a crop differ in their tolerance to surface stagnation of water. The damage may also depend upon the stage of growth at which submergence occurs as the crop may be more sensitive at one stage compared to another.

Grasses and forages are usually more tolerant to inundation as compared to cereals or other crops. In the Indian context only limited data and that too only for few crops are available in this regard (Table 4.1). With the available information to date, however, few general observations may be made. Two important observations are (i) The crops, in general, are more sensitive to water stagnation at early than later growth stages, and (ii) the yield reduction increases with increasing degree and duration of submergence. General guidelines for the periods within which stagnated water should be removed from crop lands are given in Tables 4.1 and 4.2.

Table 4.1 Permissible water stagnation period for different crops

Crop	Water stagnation period allowed (days)
Vegetables	1
Oil seeds	2
Cotton, maize	3
Rice, sugarcane	7

METHODS OF FIELD DRAINAGE

Surface drainage of agricultural lands is accomplished by several measures, such as land forming, open ditches and terracing. Land forming is defined as the process of changing the natural topography so as to control the movement of water on to or from the land surface. In the context of drainage it includes land smoothing, land grading or levelling and shallow field drains which are crossable with farm machinery. Levees commonly used for flood control, may or may not form a part of surface drainage system although many a times these are essential to control the run-off from entering the adjoining areas.

Table 4.2 Submergence tolerance of various crops at different locations in India

Site	Crop	Submergence tolerance indices		
		Threshold (days)	Slope (%)	DS_{50} (days)
Delhi	Pigeon pea	1.6	23.2	3.8
Hisar	Cowpea	0.8	6.6	8.4
	Pigon pea	0.5	9.2	6.0
Karnal	Wheat	0.0	7.0	7.2
	Pearl millet	0.0	5.3	9.4
Ludhiana	Wheat	1.9	9.2	7.3
Pusa	$Groundnut^{F}$	0.0	8.9	5.6
	$Groundnut^{P}$	0.0	10.3	4.9
	$Maize^{K}$	0.0	13.1	3.8
	$Maize^{T}$	1.2	9.6	6.4
	$Maize^{GF}$	0.2	9.4	5.5

Threshold means that there would be no reduction in the yield if submergence does not exceed the given value.
Slope means percent reduction in the yield per day of submergence beyond the threshold value.
DS_{50} means reduction in yield would be 50 % if submergence equals the number of days indicated. The yield reduction is in relation to the control where no submergence has been allowed.
Control treatment in the case of wheat crop at Ludhiana is one day submergence as against no submergence at other places
Superscripts F, P, K, T, and GF stand for flowering, pod formation, knee height, tasseling and grain formation stages, respectively.

Land Smoothening/ Land Grading

Flat lands where there are only minor differences in field levels, land smoothening is carried out to achieve good surface drainage. In this process general contours of the ground surface are not altered but only small ridges are removed and depressions filled up to achieve a smooth surface. The smoothening operation is carried out with the help of wooden floats or land planes. For best results, it is desirable to use the equipment three or more times by going across the field lengthwise, crosswise and diagonally. For successful operation, soil should be broken down to clods no bigger than the fist and should be just moist enough to crumble in the hand. Smoothening operation is also essential to finally finish up land levelling jobs which end up in a irregular land surface. This operation is quite cheap and can be carried out once a year.

Land grading and land levelling are synonymously used in surface irrigation and drainage. Land grading is more descriptive than land levelling where a sloping plane surface than a level surface is designed. Land grading is a one time operation in which pre-decided design slope is achieved. The design slope should meet the requirements of both the surface irrigation and drainage. For drainage, the grade in the direction of drains can be in the range of 0.05 to 0.5 per cent but a preferable slope is 0.1 per cent.

DRAINAGE SYSTEMS

A number of surface drainage systems are in vogue. Some of them are bedding system, parallel field drain system, the random drain system and the parallel open ditch system.

Bedding System

Bedding system is probably the oldest of the surface drainage systems. It is suitable for soils with poor hydraulic conductivity with slopes not exceeding 1.5 per cent. The level surface of the field is converted into a sloping area by ploughing several times in such a way as to always throw the back furrow in the channel. The beds are always kept parallel to the greatest slope of the field such that the dead furrow serves as a field drain. The width of the bed depends upon several factors including the slope, the land use, soil permeability, farm machinery to be used and the nature of the farming operations. Based on hydraulic conductivity, the bed width varies from 8 to 30 m (Table 4.3).

Table 4.3 Bed widths in relation to hydraulic conductivity of soils

Hydraulic conductivity (m/day)	Bed width (m)
$K \leq 0.5$	8-12
$0.05 < K \leq 0.10$	15-17
$0.10 < K \leq 0.20$	20-30

The length of the beds could be in the range of 100 to 300 m. The height of the crown from the bottom of the furrow should not exceed 45 cm for pastures and 20 cm on arable lands. Field laterals are provided perpendicular to the field drains which collect the water from these drains and carry it to the main drain. The bedding system has many disadvantages particularly for row crops. When rows are parallel to the dead furrows, the flow to the furrows is obstructed. Moreover, the slope of the furrow may not always be sufficient to take care of the overland flow. The system also requires movement of the top soil from the side of the beds to the middle and therefore, fertility gradients could be established at least in the initial years. Restrictions in the mechanized farming could be another limitation of this method. Maintenance costs are also high as the beds would require regular maintenance. The method, however, is most useful for pasture lands.

Parallel Field Drain System

The parallel system is used when the surface is uniform and has few noticeable ridges or depressions. The channels in this case are spaced farther apart than a bedding system and have greater capacity than the dead furrows. The design and layout is similar to that of bedding system but in this case it is not essential that drains are spaced equally or they drain on both sides. The depth of the drain should be a minimum of 20 cm and have a minimum cross section of 0.45 sq.m. The side slopes are 8:1 or flatter. The row or field drains are perpendicular to the field lateral. The system is most suitable for fields where agricultural operations are mechanized. Land smoothening of field surface between laterals on regular basis is essential. It will keep the land well graded. It would also improve the efficiency of the drainage system.

Random System

These drains are best suited to the drainage of scattered depressions. Field or row drains and field laterals meander from one depression to another to drain the same to the outlet ditches. In this system, location of the lateral ditches is dictated by the topography of the land. The system is most suitable in areas where depressions are deep and it is technically or economically not feasible to fill up the depressions. The slope of the channel can be as flat as 10:1 when depth of drains is more than 60 cm or 4:1 if fields are farmed parallel to the drains. The minimum cross section area of 0.45 sq.m. is recommended. The channels should follow a route that requires minimum cut. While aligning the channels, it should be ensured that there is minimum interference in the farming operations. The system costs much less and provides more returns on investments. The maintenance cost is relatively more than other systems of drainage.

Parallel Lateral Open Ditch System

The parallel lateral open ditch is a system which provides surface drainage to the area as well as helps to control the ground water table. The layout of this system is similar to the field drain system but differs in a way that ditches are deeper in this case. When financial or other constraints do not allow to lay tile (pipe) drainage, this system of drainage usually referred to as open drainage system is commonly recommended till finances are generated or made available to lay the pipe lines.

HORIZONTAL SUBSURFACE DRAINAGE

In order that a high level of productivity could be developed and sustained, agricultural lands in the humid and sub-humid regions and irrigated lands generally need subsurface drainage. The challenge is to correctly evaluate the problem, design and install most economical drainage system and to maintain the system so that maximum benefits accrue from the land.

A horizontal drainage system consists of the laterals, the collectors and the main drains. Lateral drains (Fig. 4.1), laid parallel to each other to cover the entire field, accept the excess ground water from the soil above the drain level and hence serve primarily to control the ground water table. The water entering into the laterals flows into the collectors which carry it to the main drain. The main drains carry the water to the outlet from where it is disposed out of the area either by gravity or by pumping from a sump.

Figure 4.1 A view of the lateral field drain

Drainage Investigations

In order to determine the feasibility of the drainage and for the design and installation of an efficient and economical drainage system, investigation of site conditions is essential. Surveys and investigations for drainage projects should be carried out to reveal the following information :

i. Inventory of the project site conditions

ii. Causes of the excess water

iii. Identification of appropriate solution

iv. Data needed to establish drainage system design parameters
   a. Topographic map
   b. Data on soil salinity/alkalinity, soil profile description and depth to impermeable layer etc.
   c. Position and fluctuations of the ground water table, depth to water table, artisan pressure and ground water quality
   d. Hydraulic conductivity/drainable pore space
   e. Crops and their tolerance to waterlogging
   f. Irrigation practices and water requirement

The information could be obtained by systematic surveys and investigations. The amount of investigations required would mainly depend upon the investigator's knowledge of the area, manpower and the funds at disposal, amount of data already in hand or available with different agencies operating in the area, size of the project and data available for similar projects executed earlier. Projects covering large areas would require more intensive investigations.

Water Transmission Characteristics

The two most important soil characteristics that are commonly used and referred to in a drainage investigation and design are : (i) the hydraulic conductivity, and (ii) the drainable pore space. Reliable estimates of these soil characteristics are the backbone of a pipe drainage system.

There are several methods of determining the hydraulic conductivity. A first guess of the hydraulic conductivity can often be made on the basis of soil texture (Table 4.4). For drainage design, invariably *in situ* determination of hydraulic conductivity is desirable. The *inverse auger hole* method can be employed to determine the horizontal hydraulic conductivity of soil layers above the water table, while the *auger hole* method is recommended for soil layers below the water table. The auger hole method is applicable for both homogeneous and layered soils provided the pertinent layers are 10 times or more thicker than the radius of the auger hole. The method is ideal for a depth upto 5 m. Piezometer method is used to determine the hydraulic conductivity of soil layers as thin as 10 cm and for depths upto 8 m. Readers are advised to refer a standard text book on drainage for more information on conducting *in situ* hydraulic conductivity evaluation tests.

Table 4.4 Relationship between hydraulic conductivity and soil texture

Soil texture	Hydraulic conductivity (cm/hr)
Clay	.0025 - .0380
Silty clay	.0381 - .0635
Silty clay loam	.0635 - .1016
Clay loam	.1016 - .1778
Sandy clay loam	.1778 - .3810
Silt loam	.3810 - .8890
Loam	.8890 -1.7780
Very fine sandy loam	1.778 - 3.048
Fine sandy loam	3.048 - 4.318
Sandy loam	4.318 - 6.350
Loamy fine sand	6.350 - 8.890
Loamy sand	8.890 -11.430
Fine sand	11.430 -15.240
Sand	15.240 -20.320
Coarse sand	20.320 -30.480

Drainable Porosity

The drainable porosity also known as effective porosity and specific yield is defined as the volume of water released or taken into storage in an unconfined aquifer per unit cross sectional area per unit decline or rise of water table as a fraction of total volume of the soil. For drainage conditions, drainable porosity of a soil relates to the amount of fluctuation of the water table to the amount of ground water added or drained from the system. To determine the build up of the water table from each increment of recharge, the depth of each recharge should be divided by the drainable porosity.

Both laboratory and field determinations of drainable porosity are expensive and time consuming. A large number of tests are needed to obtain a representative drainable porosity of the area. On the basis of considerable data, a general relationship (Fig. 4.2) has been developed between the hydraulic conductivity and drainable porosity (US Bureau of Reclamation, 1971). Since the fluctuation of the water table in a drained area takes place in the zone between the drains and the maximum allowable water table height, the drainable porosity corresponding to the weighted average hydraulic conductivity of this zone can be used in drain spacing calculations in most cases.

Another approach to estimate drainable pore space is on the basis of soil texture and soil structure which have considerable influence on the drainable pore space. The representative volumetric drainable porosity values for different textures of soil can be taken as 3-5 % for heavy clays, 6-9 % for medium textured soils and 10-15 % for sandy soils. Exact estimates can be made from the data presented in Table 4.5. Whenever possible, the drainable porosity should be determined from the measurements of drain discharge and drawdown of existing or pilot drains.

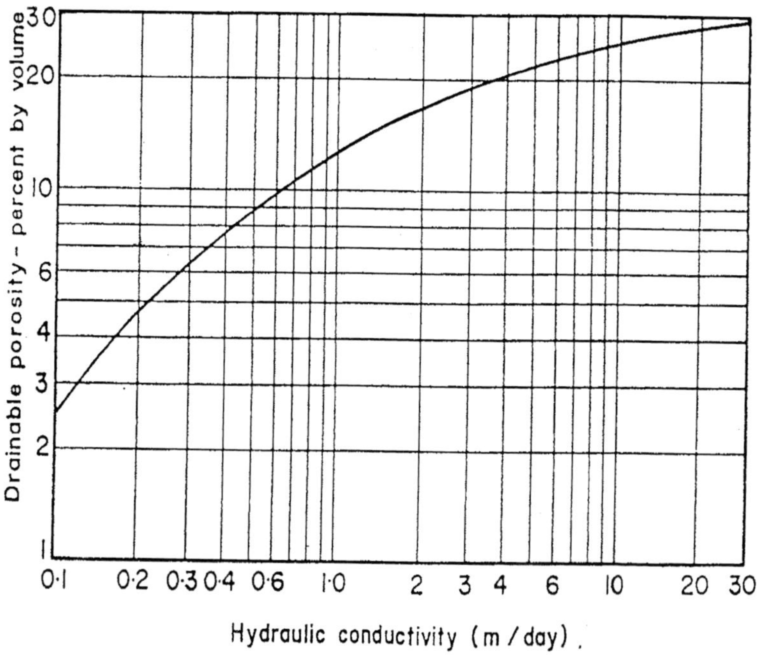

Figure 4.2 Relationship between hydraulic conductivity and drainable pore space

Table 4.5 Influence of soil texture and structure on drainable pore space

Texture	Structure	Drainable pore space (per cent)
Clay Heavy clay loam	Massive, very fine or fine columnar	1 – 2
Clay Clay loam Silty clay Sandy clay loam	Very fine or fine prismatic, angular blocky or platy	1 – 3
Clay Silty clay Sandy clay Silty clay loam Clay loam Silty loam Silt Sandy clay loam	Fine and medium prismatic, angular blocky and platy	3.8
Light clay loam Silt Silt loam Very fine sandy loam Loam	Medium prismatic and sub-angular blocky	6–12
Fine sandy loam Sandy loam	Coarse sub-angular blocky and granular, fine crumb	12 – 18
Loamy sand Fine sand	Medium crumb and single grain	15 – 22
Medium sand	Single grain	22 – 26
Coarse sand Gravel	Single grain	26 – 35

Based on the data from the Water and Power Resources Survey (Formerly United States Bureau of Reclamation, 1971).

DRAINAGE COEFFICIENT

The drainage coefficient of a drainage system needs to be determined accurately because it controls the drain spacing, the size of the field laterals, and the capacity of the collectors and pumping station (if any). Excess rain or irrigation water or net ground water inflow, that causes rise in the water table to a level shallower than the permissible limits, must be drained out. Although elaborate procedures are used in evaluating drainage coefficients, general guidelines for selecting drainage coefficient based on the rainfall are available for the United States of America (Table 4.6). The range of drainage coefficients assessed/observed at few places in the country are also reported in Table 4.7. These figures could form the guidelines in selecting drainage coefficients for arid and semi-arid regions.

It is seen that subsurface drainage coefficients in our country are on the lower side. It is because, in India, most of the rains occur in a few months period generating larger amounts of run-off.

Table 4.6 Subsurface drainage coefficients for different rainfall regions based on data from USA

Rainfall (mm)	Drainage coefficient (mm/day)
< 762	6.35 – 7.87
762 – 1016	7.87 – 9.65
1016 – 1270	9.65 – 12.70
1270 – 1524	12.70 – 25.40
> 1524	≥ 25.40

Table 4.7 Subsurface drainage coefficients from different locations (observed from test plots)

Site	Rainfall (mm)	Rate (mm/day)	Recommended range (mm/day)
Chambal (Rajasthan)	850*	3.0	2.5 – 3.5
Sampla (Haryana)	600	2.5	2.0 – 3.0
Hisar (Haryana)	400	2.0	1.5 – 2.5
Dabhou (Gujarat)	800	4.0	3.0 – 5.0
Mundlana (Haryana)	500	5.0	5.0 – 7.0
Kailanakhas(Haryana)	500	6.8	5.0 – 7.0
Muraj (Gujarat)	500	2.8	2.0 – 4.0

* Variation in the range of 600-1400 mm.

DESIGN OF SUBSURFACE DRAINAGE SYSTEM

In the design of subsurface drainage system the following main items have to be considered.

i. Laterals: Depth and spacing of field laterals

ii. Collector: Diameter and slope for carrying capacity

iii. Sump

iv. Pumping capacity

Laterals: Depth of Drains

Important aspects to be considered while determining depths for the subsurface drains are:

i. The desirable depth to water table for proper aeration in the root zone.

ii. The depth of drain placement (it should be such that the cost per hectare is minimum (Fig. 4.3)

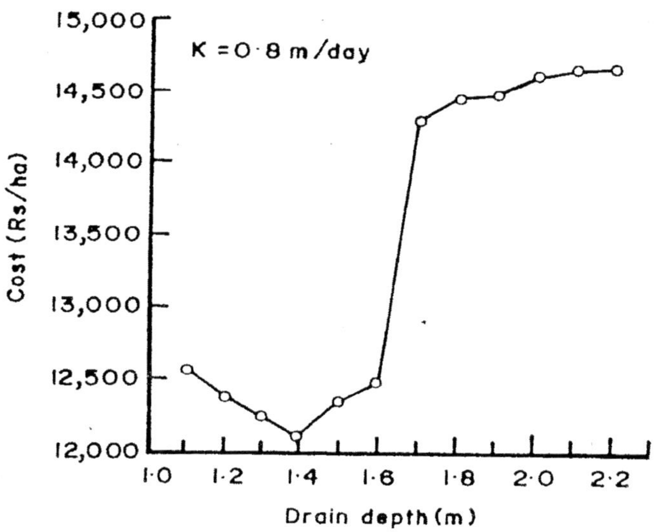

Figure 4.3 Drain depth versus cost of drainage

iii. Soil conditions such as unstable or hard soils and quick sand conditions.

iv. The available depth of outlet drains, topography, minimum pipe slopes.

For irrigated annual crops in loam, sandy loam and silt loam soils, average water table depth of 0.60 to 0.80 m can be tolerated. For clay loam and clay soils, this depth is 0.8 to 1.0m. Perennial crops and especially trees need deeper depths to water table. The recommended depths to water table for various crops are listed in Table 4.8. An average seasonal water table depth of about 1.0 m will provide reasonably favourable conditions in the root zone for most of the crops.

The depth of the drains should be such that midway between the drains, the water table should be at a satisfactory depth during the critical period. It means, if a crop requires a depth to water table of 1m then the water table at midpoint between the drains be at least 1 m deep during the cropping season.

Table 4.8 Optimum depth of water table at different locations

Crops	Depth of water table (cm)
Pearl millet	125
Wheat	30-50
Peas	50
Maize (Corn)	60-100
Soybean	125-150
Cow pea	75
Sorghum	75
Barley	100
Mustard	125
Gram	100
Sugarcane	75
Safflower	100
Groundnut (summer)	100
Groundnut (winter)	50
Rice	< 30
Green gram	60

In practice drain depth varies greatly. The minimum and maximum values are commonly 50cm and 2m, respectively. The minimum recommended soil cover should always be provided to avoid damage due to continuous use of the heavy machinery (Table 4.9). From practical view point, a depth of 1.2 to 1.8m is optimum for salinity control. The other factors which control the depth of lateral drains are described below.

Table 4.9 Minimum recommended soil cover over tiles

Soil type	Depth of soil cover (cm)
Mineral	60
Deep peat and Muck	120
Organic	75

Soil Salinisation and Drain Depth

There are two schools of opinion on this subject. While one advocates shallow drains the other is in favour of deep drains. The advocates of deep drains usually rely on the concept of critical depth to water table arguing that it avoids secondary salinisation in addition to several other advantages which might accrue on this account. For example, deeper drains will allow larger lateral drain spacing. On the other hand, recent studies have conclusively proved that secondary salinisation is often avoidable in case flux at the water table could be kept downward. Therefore, shallow drains could be as effective as deeper drains provided it is ensured that at critical periods flux is downward. In monsoon climatic regions, when the shallow water table conditions prevail, the flux on an average would be downward. Therefore, in the Indian context, shallow drains (depth 1.2 - 1.5m) could be effective in reclamation and management of saline soils.

Shallow versus Deep Drains

The choice of the type of drainage may depend on whether a deep or shallow drainage system will be applied. The deep drains have the advantages of smaller loss of land, wider spacing, larger open water and soil storage. Contrary to advantages, a deep drain system will have higher installation costs, both with respect to the higher excavation costs and the additional costs for larger pipes in the case of covered drains. An analysis made at CSSRI, Karnal on the cost of the materials and manual installation of pipe drains showed that the cost is significantly affected by the choice of the drain depth. The drain depth of 1.4 m and the corresponding drain spacing results in the lowest total cost for providing drainage system in sandy loam soils (Fig. 4.3). Sometimes, maintenance also may be a limiting factor in deciding drain depth. In case of pumped outlets, extra cost of pumping due to larger elevation must also be taken into account.

Spacing of Lateral Drains

The water table between two parallel drains is usually curved (Fig. 4.4), its elevation being highest midway between the drains. The factors which influence the height of water table are:

— Precipitation and other sources of recharge
— evaporation and other sources of discharge
— soil properties
— depth and spacing of the drains
— cross-sectional area of the drains
— water level in the drains

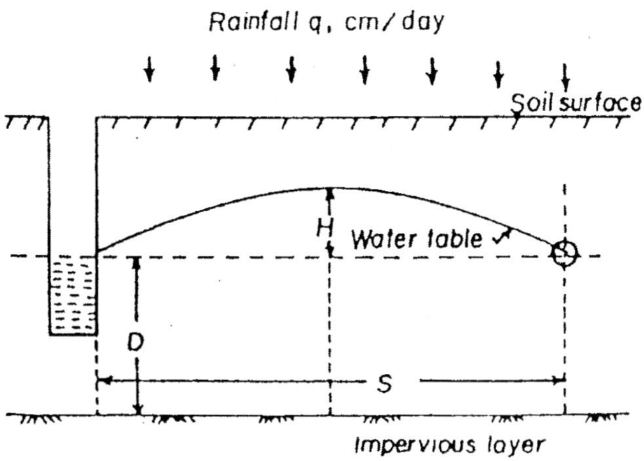

Fig. 4.4 Steady state water table position in drained lands

Several equations were developed interrelating the above factors. These equations are used to calculate the lateral drain spacing.

The steady state formulae are derived with the assumption that the recharge intensity equals the drain discharge rate and as a consequence the water table remains at a constant position. In practice, the average water table and discharge of the season are taken for computations. The most commonly used steady state equation for determining drain spacing is the Hooghoudt's equation, which can be written as below: (Hooghoudt, 1940)

$$q = \frac{8K_b\ DH + K_a\ h^2}{L^2} \tag{4.2}$$

where,

q = drain discharge rate per unit surface area (m/day) in relation to a recharge rate such that recharge is equal to the discharge rate;

h = water table height above drains at mid point between drains (m);

D = height of the water level in the drains above the impervious floor i.e. thickness of aquifer below drain level (m);

L = drain spacing (m);

K_a = hydraulic conductivity of the layer above drain level (m/day) and

K_b = hydraulic conductivity of the layer below drain level (m/day).

If the drains do not reach the impervious floor, the flow lines will not be parallel and horizontal but will converge towards the drain (radial flow) causing lengthening of the flow lines and consequently more than proportional loss of head. As a result, the elevation of the water table will be higher where the vertically walled drains are replaced by pipe drains, the drain level remaining the same. To account for the extra resistance caused by radial flow, Hooghoudt introduced a reduction in the depth D to a smaller equivalent depth d. The factor d is a function of L, D and r_o (radius of drain) and the values of d for various values of L and D and r_o can be calculated on the basis of following equations :

$$d = \frac{D}{1 + D/L\,(8/\pi \ln (D/r_e) - \alpha)} \quad ; 0 < D/L \le 0.3 \qquad (4.3)$$

and

$$d = \frac{\pi D}{8\,[\ln (D/r_e) - 1.15]} \quad ; D/L > 0.3 \qquad (4.4)$$

where,

$$\alpha = 3.55 - 1.6D/L + 2(D/L)^2 \qquad (4.5)$$

and

$$r_e = r_o + \text{fthick} \qquad (4.6)$$

r_e is the effective drain pipe radius and fthick is the filter thickness.

Eq.(4.2), by replacing D with d, can be rewritten as:

$$q = \frac{8K_b\ dh + 4\ K_a\ h^2}{L^2} \qquad (4.7)$$

This equation is commonly used to calculate the drain spacing L, if other factors are known. As d is a function of L, the use of the formula involves a trial and error procedure. To avoid the trial and error, a number of nomographs have been prepared so that drain spacing could be calculated with minimum complexity. Even then the availability and application of nomographs have been very limited. Presently, with availability of table top computers, it is easy to calculate drain spacing using simple computer programmes than to depend on the nomographs. A computer programme for solving the Hooghoudt equation is as follows:

Computer Programme to Solve Hooghoudt Equation (Eq. 4.7)

```
      REAL KAVG,K1,K2
      CALL FILEI
      CALL FILEO
      READ(5,*)NC
2     READ(5,*)N,Q,RAD,HMID,FTHICK,K1,K2,D1,DIMP
      WRITE(6,5)N
      REQ=RAD+FTHICK
      PI=3.1415926
```

```
                    *DAGAN EQUATION
      BETA=2.0/PI*ALOG(2.0*COSH(PI*RAD/DIMP)-2.0)
      C=-8.0*K2*HMID*DIMP/Q
      B=-2.0*BETA*DIMP
      SDAG=(-B+SQRT(B*B-4.0*C))/2.0
                    *HOOGHOUDT EQUATION
      SDRAIN=SDAG
80    DIMD=DIMP/SDRAIN
      IF(DIMD.LE.0.3)THEN
      ALPHA=3.55-(1.6*DIMD)+2*DIMD**2
      DEQ1=DIMP/(1.0+DIMD*(8.0/PI*ALOG(DIMP/REQ)-ALPHA)
      ELSE
      DEQ1=SDRAIN*PI/(8.0*(ALOG(SDRAIN/REQ-1.15))
      ENDIF
      SHOOG=SQRT((8.0*K2*DEQ1*HMID+4.0*K1*HMID**2)/Q)
      IF(SDRAIN.LT.(0.9*SHOOG))GOTO 100
      IF(SDRAIN.GT.(1.1*SHOOG))GOTO 110
      GOTO 120
100   SDRAIN=SDRAIN+200.0
      GOTO 80
110   SDRAIN=SDRAIN-200.0
      GOTO 80
120   IF(SDRAIN.LT.(0.95*SHOOG))GOTO 130
      IF(SRAIN.GT.(1.05*SHOOG))GOTO 140
      GOTO 150
130   SDRAIN=SDRAIN+50.0
      GOTO 80
140   SDRAIN=SDRAIN-50.0
      GOTO 80
150   WRITE(6,800)SHOOG
      IF(N.LT.NC) GOTO 2
      CONTINUE
5     FORMAT(//1X,'EXAMPLE',I3)
800   FORMAT(/1X,'SPACING BY HOOGHOUDT EQU.=',F9.2)
900   END
      SUBROUTINE FILEO
```

```
      CHARACTER FNAME*40
      WRITE(*,11)
11    FORMAT(5X,' TYPE THE NAME OF OUTPUT FILE ')
      READ(*,10)FNAME
      OPEN(6,FILE=FNAME )
10    FORMAT(A40)
      END
      SUBROUTINE FILEI
      CHARACTER FNAME*40
      WRITE(*,11)
11    FORMAT(5X,' TYPE THE NAME OF INPUT FILE     ')
      READ(*,10)FNAME
      OPEN(5,FILE=FNAME )
10    FORMAT(A40)
      END
```

The input file is prepared in the free format and output is received as the spacing as calculated by the Hooghoudt equation (Annexure II).

Non Steady State Solutions

Most commonly used non-steady state solution for designing lateral drain spacing in cases where drains do not rest on the impermeable layer is that of Glover as reported by Dumm (1954).

The schematic view of the water table under non-steady state is shown in Fig. 4.5. The solution for homogeneous soils and for no recharge is given as,

$$L = \pi \left[\frac{KDt}{f} \right]^{1/2} \left[\ln 1.27 \frac{h_o}{h_t} \right]^{1/2} \tag{4.8}$$

Here K is the hydraulic conductivity of the soil, f is the drainable pore space, t is the time to bring water table height above the drain level from an initial level of h_o to the desired value h_t . The equation is valid for d >> h_o.

Dumm (1964) argued that in drained lands water table initially would not be flat as assumed in deriving eq. (4.8). Assuming the shape of the water table initially as a fourth degree parabola, a drainage design equation which is very

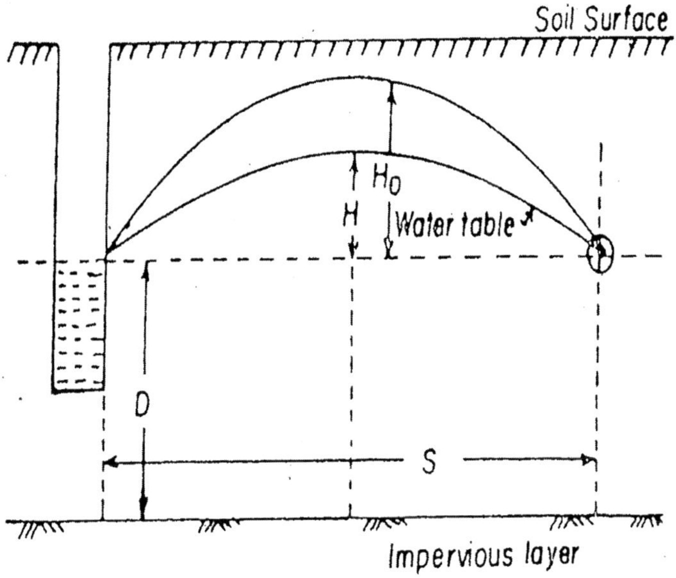

Fig. 4.5 Water table drawdown in drained lands as anticipated in the derivation of non-steady state equations

similar to eq. (4.8) has been derived. It differs only in the value of the constant 1.27 which is now replaced with 1.16. The equation is known as the USBR (United States Bureau of Reclamation) equation and is very commonly used for drainage design in the USA. The exact equation is written as,

$$S = \pi \left[\frac{KDt}{f}\right]^{1/2} \left[\ln 1.16 \frac{h_o}{h_t}\right]^{-1/2} \tag{4.9}$$

Glover-Dumm eq. (4.8) and eq.(4.9) do not take into account the radial resistance of the flow towards drains not reaching an impervious layer. To account for the convergence of the flow in the vicinity of the drains, the thickness of the aquifer, D is replaced by d as in the case of the Hooghoudt equation. With this change, eq.(4.9) is written as follows.

$$S = \pi \left[\frac{Kdt}{f}\right]^{1/2} \left[\ln 1.16 \frac{h_o}{h_t}\right]^{-1/2} \tag{4.10}$$

Another modification to eq.(4.9) has been suggested by Tapp and Moody (Dumm, 1964), to take care of flow from the region above the drains. According to this modification, d in eq.(4.9) should be taken as $(d + h_o/2)$ or the average initial depth of water bearing stratum.

In using Glover equation or the USBR equation for drainage design, a design criterion is first decided. The usual design criterion for the Indian condition is to lower the water table by 30 cm in 2 days time once the water table reaches the soil surface. Other similar criterion may be developed for a particular site and cropping conditions. A computer programme which solves the USBR equation is given as follows:

Computer Programme to Solve USBR Equation (Eq. 4.10)

```
TO SOLVE USBR EQUATION
NC=NUMBER OF CASES TO BE SOLVED
N = CASE NUMBER
RAD=RADIUS OF THE DRAIN,CM
FTHICK=TICKNESS OF THE FILTER MATERIAL,CM
KAVG=HYDRAUIC CONDUCTIVITY, CM/DAY
D1 = DEPTH OF DRAINS,CM
DIMP=DEPTH TO IMPERMEABLE LAYER BELOW THE DRAINS,
        CM
T= TIME TO LOWER THE WATER TABLE FROM D1 TO HT, DAYS
DPOR = DRAINABLE PORE SPACE,FRACTION
HT= HEIGHT OF THE WATER TABLE ABOVE DRAIND AFTER
     TIME T,CM
```

```
      REAL KAVG
      CALL FILEI
      CALL FILEO
      READ(5,*)NC
2     READ(5,*)N,RAD,FTHICK,KAVG,D1,DIMP,T,DPOR,HT
      WRITE(6,5)N
      REQ=RAD+FTHICK
      PI=3.1415926
                *USBR EQUATION
      SDRAIN=40
170   DIMD=DIMP/SDRAIN
      IF(DIMD.LE.0.3)THEN
```

```
      ALPHA=3.55-(1.6*DIMD)+2*DIMD**2
      DEQ2=DIMP/(1.0+DIMD*(8.0/PI*ALOG(DIMP/REQ)-ALPHA)
      ELSE
      DEQ2=SDRAIN*PI/(8.0*(ALOG(SDRAIN/REQ-1.15)))
      ENDIF
      SGDUM=PI*SQRT(KAVG*DEQ2*T/DPOR)/SQRT(ALOG
         (1.16*D1/HT))
      Y1=ABS(SDRAIN-SGDUM)
      IF(Y1.LE. 0.5) THEN
      GOTO 240
      ENDIF
      SDRAIN=(SDRAIN+SGDUM)/2
      GOTO 170
240   WRITE(6,860)SGDUM
      IF(N.LT.NC) GOTO 2
      CONTINUE
5     FORMAT(//1X,'EXAMPLE',I3)
860   FORMAT(/1X, 'SPACING BY GLOVER-DUMM EQU.=', F9.2)
900   END
      SUBROUTINE FILEO
      CHARACTER FNAME*40
      WRITE(*,11)
11    FORMAT(5X, 'TYPE THE NAME OF OUTPUT FILE')
      READ(*,10)FNAME
      OPEN(6,FILE=FNAME )
10    FORMAT(A40)
      END
      SUBROUTINE FILEI
      CHARACTER FNAME*40
      WRITE(*,11)
11    FORMAT(5X, 'TYPE THE NAME OF INPUT FILE')
      READ(*,10) FNAME
      OPEN(5,FILE=FNAME )
10    FORMAT(A40)
      END
```

The input file in this case is also prepared in the free format and output is received as the spacing as calculated by the USBR equation (Annexure II).

In addition to the demands of watertable and salinity control, some other factors may also play a vital role in deciding drain spacing. For example, the Haryana farmers prefer to have drains on their farm boundaries. After the land consolidation, the farms attained a regular rectangular shape of 60 x 67 m. Therefore the drain spacings are also either 60 or 67 m depending on the direction of the drains. Likewise in practice, the area will be divided into 'blocks of equal spacings' and the calculated values will be rounded off to the nearest figures of 20 m, 25 m, 30 m, 35 m, 40 m, 45 m, 50 m, 55 m, etc.

Depending on experimental results, many countries have standardized the design, drain depths and spacings as per the soil conditions. Table 4.10 gives the practices followed in U.S.A. alongwith some experimental results from India.

Size of Lateral Drain Pipe

Wessling's equations for uniform flow in smooth pipes and corrugated pipes derived from Manning equation can be used to calculate the size of the lateral drain pipes. The equations are:

Smooth pipes : $Q = 89\ d^{2.714}\ i.^{0.57}$ (4.11)

Corrugated pipes : $Q = 38\ d^{2.67}\ i^{0.5}$ (4.12)

Where, Q = discharge along the pipe = q A, m^3/sec

d = internal diameter of pipe, m

i = hydraulic gradient, m/m (Table 4.11)

Normal practice is to 'over design', the pipe to allow for subsequent partial silting and for misalignment during construction. This is possible by increasing the discharge Q by a factor varying from 1.33 for continuous plastic pipes installed in stable soil to 2.0 for the tile drains laid in situations where it is feared that siltation may occur. In case of clay and cement concrete tiles the minimum diameter prescribed is 10 cm, while for PVC and PE pipes, it is 7.5 cm. The computed pipe diameters are to be rounded off to the next higher commercially available size.

Table 4. 10 Average depth and spacing of tile drains (U.S.D.A. S.C.S)

Soils	Hydrology class	Conductivity rate (m/day)	Spacing (m)	Depth (m)
USDA–SCS				
Clay	very slow	< 0.03	9 - 15	0.9 - 1.1
Clay loam	slow	0.03-0.1	12 - 21	0.9 - 1.1
Loam	moderately slow	0.1 - 0.5	18- 31	1.1 - 1.2
Sandy loam	moderate	0.5 - 1.5	31 - 37	1.2 - 1.4
	moderately rapid	1.5 - 3.1	31 - 61	1.2 - 1.5
Peat and Muck	rapid	3.1 - 6.1	31 - 92	1.2 - 1.5
Irrigated soils	variable	–	46-183	1.5 - 2.4
India				
Sandy loam (Haryana)		0.8 - 1.5	65 - 75	1.4 - 1.7
Loam (Gujarat)		0.3 - 0.5	50	1.5

Table 4.11 Minimum grade, i for tube drain

Drain size (mm)	Grade (%)
100	0.10
125	0.07
150	0.05

Drainage Sump

In India where monsoon type of climate is experienced, the ideal period for construction of a sump is May or June when the groundwater is deepest. A circular pit of diameter 1 m more than the diameter of the proposed sump is dug up to the ground water level. At the bottom of the circular pit an iron ring is placed to serve as cutting edge over which a RCC kerb is constructed to support the masonry wall to a height 1m above the ground. The wall is plastered from inside and outside and is provided with holes of 1 cm diameter at regular intervals. The sump is sunk into the ground by removing the soil from the bottom in case shallow water table conditions prevail. Simultaneously, the wall of the sump is also raised such that the sump is 2 to 2.5m deeper than the depth at which collector drain has been laid. In waterlogged sandy soils, quick sand conditions may interfere with the installation of sump. Such situations can be tackled by simultaneously dewatering the area all around the sump while sinking the structure.

On completion of the construction of the sump, a gravel layer of about 0.5 m thickness is placed at the bottom of the sump to serve as a filter for the ground water flowing into the sump. Holes are provided in the side walls for connecting the collector drains. A sump of 3.5 m to 4 m inner diameter and of 4 m depth below the ground could serve an area of about 40 ha.

INSTALLATION OF SUBSURFACE DRAINAGE SYSTEM

Based on the field investigations and crop drainage requirements, a drainage system is designed and a layout plan is prepared to suit the field boundaries. The topography of the project site receives due weighage in deciding the layout plan. For the installation of drainage system the collector and lateral drain lines are staked and the location for outlet and junction boxes are identified in the field. The installation procedures of the two main components of the horizontal subsurface drainage system viz. the outlet, the collector and lateral drains are as follows:

Outlet

The execution of drainage system begins with the construction of the outlet which could either be a gravity or a pumped outlet. If the natural slope permits the disposal of drainage water from the collector or main drain into an open water body should be considered. The gravity outlets are preferred over the pumped outlets because these are cheaper to construct and do not involve operational cost.

In flat or gentle-sloping areas, the gradient is generally insufficient for free fall, and disposal of water through gravity into the open watercourse is difficult. Under these conditions, a sump is constructed in between the subsurface collector drain and the open channel. The sump provides space for storage of water flowing from the collector drain, before it is pumped into the open watercourse.

Collector and Lateral Drains

According to the drainage plan, the location of the collector and lateral drains is first marked on the ground. The construction work is initiated during a part of the year when the water table is deep enough to facilitate the earth work. The digging of trenches begins from the outlets, so that these can be kept free of water during rains etc. The trenches for the collector drains should be prepared first. After preparing the trenches for the collector drain, the bed slope is thoroughly checked for any reverse slope. Rigid PVC pipes, stone-ware pipes, or perforated corrugated pipes can be used for the collector drains. Perforated corrugated pipes, if used, need to be provided with a suitable envelope material in sandy and loamy soils. Water in the trenches if any is removed through the outlet or pumped out through the sump. It is particularly essential when rigid PVC pipes are used because if drains are laid in ponded water condition the pipes will float on the water. In such conditions, it would be difficult to maintain the desired slope.

For lateral drains, once the bed slope is ready, the gravel envelope is spread in the 30 cm strip at the bottom before the pipe is laid (Fig. 4.6). The pipe is then covered with 10 cm thick gravel layer. Process of putting the envelope into the trench and above the drains is quite important as it could make or mar the performance of the drainage system. The drainage pipes with the envelope need to be set in a stabilized soil surrounding. For good performance, the pipes should be laid in such a way that soil slurry is not formed. High water table during installation of drains also adversely affects the performance of the envelope as sharp hydraulic gradients are developed resulting in high inflow velocities of water which carry the soil particles into the drain. When installation is made under such conditions, the amount of silt carried into the drain may be tolerable to some extent if no filter has been used. With a filter wrap, even a short term exposure to high water table conditions can effectively clog the openings of the filter. Once the drains become operational there is little chance of building up of high hydraulic gradients. The chances of sediments flowing into the drains are maximum during installation and/or during first leaching phase.

Figure 4.6 Laying a lateral drain with gravel envelope.

REUSE\DISPOSAL OF DRAINAGE WATER

One of the major issues that confronts a drainage engineer is safe disposal of the drainage water that is received through the drainage system. The disposal and/or its reuse must ensure that there are no adverse effects on the soil and water resources of the region and the environment at the disposal site remains undisturbed. Both the quality and the quantity of drainage water would influence the selection of safe means of reuse/disposal. One of the plus feature of subsurface horizontal drainage over the vertical drainage system is that in the former, quantity of drainage water is much less and therefore the disposal problems would be minimal. Secondly, the quality of drainage water in the deeper strata is usually poor. Thus, pipe drainage would offer more opportunities for the reuse of drainage water compared to the tubewell drainage. The temporal variation in the quality of drainage water from a pipe drainage system, the quality of water from a cavity well (20 m depth) and the canal water at the experimental site at Sampla generally attest to this trend (Table 4.12). It also reveals that there is significant improvement in the quality of drainage water over time.

Table 4.12 Average composition of drainage water, ground water and the canal water during irrigation at Sampla

Water	EC (dS/m)	SAR	Solute concentration (meq/L)			
			Ca+Mg	Na	HCO_3	Cl
Drainage water						
1986-87	26.5	16.8	173	156	1.8	315
1987-88	27.0	17.0	187	164	2.2	325
1988-89	18.0	14.7	135	121	2.1	232
1989-90	15.0	14.5	126	115	1.6	221
Ground water (20m)	32.5	11.9	280	141	4.7	405
Canal water	0.7	0.8	2	0.8	2.0	0.8

Reuse Strategies

There are three distinct possibilities for the reuse of saline drainage waters for irrigation.

i. Direct use of saline drainage water

ii. Mixing of saline drainage water with canal water

iii. Cyclic or rotational use of saline drainage water with canal water.

Considerable work has been done in the country for identifying critical limits for use of saline water for crop rotations prevalent in different agro-climatic regions. The limits for the use of saline water for different crops in different textural soils are summarized in Table 4.13 (Gupta, 1979). The experimental evidence generated in the All India Co-ordinated project on the Use of Saline Water in Agriculture, has revealed an immense potential of use of saline waters for crop production (Gupta, 1990). Data on threshold tolerance and the slope which defines the yield decrease per unit increase in irrigation water salinity over and above the threshold are reported in Annexure III (Mishra et al., 1992). Reuse of saline drainage water for irrigation eventually increases the salinity of drainage discharge and limits its further reuse for arable cropping. Studies conducted towards using such waters in agroforestry have resulted in identifying salt tolerant trees like *Casuarina*, *Tamarix* and *Prosopis* which can be grown with a modest amount of saline drainage water considered unsuitable for agricultural crops. Salt tolerant trees and halophytic shrubs and other plants have the potential to utilize drainage water while producing biomass for different purposes.

The exact mode of application of drainage water for irrigation depends on the quality of water, soil type, crops to be grown and the agro-climatic conditions of the area. While relatively lesser saline waters could be directly used for crop production, more saline waters need to be applied in conjunction with fresh water either by mixing or in cyclic application mode. In the mixing mode fresh water is used to dilute the saline water to bring its quality within permissible limits. On the other hand in the cyclic mode two sources of water are separately used in conjunction with each other without mixing.

Table 4.13 Guidelines for using saline waters for different crops (rainfall zone 400-500 mm)

Soil texture	Crop	Upper limit	
(% clay)	tolerance	EC (dS/m)	SAR
Fine	S*	< 1.5	< 5
(>30)	ST	1.5	5
	T	3.0	10
Moderately Fine	S	1.5	5
(20-30)	ST	3.0	10
	T	5.0	20
Moderately	S	3.0	10
coarse	ST	5.0	20
(10-20)	T	10.0	30
Coarse	S	5.0	20
(<10)	ST	10.0	30
	T	> 10.0	> 30

* S-Sensitive, ST-Semi-tolerant, T-Tolerant.

Cyclic use also known as sequential application or rotational mode is a technique in which canal water is replaced with saline drainage water either in a pre-decided sequence/ cycle or in switching mode. Switching mode implies that after one or more irrigations with canal water, subsequent irrigations are given with saline drainage water. The timing and amount of substitution will change with the quality of the drainage water, soil type, the cropping sequence, the climate and the method of irrigation. With cyclic use of canal and saline drainage waters, high crop yields could be obtained because maximum root zone salinity due to continued use of saline drainage water will not be experienced by the plants when such waters are used for a limited number of irrigations only. It is now known that cyclic strategy is more flexible and has many advantages over the mixing mode (Minhas and Gupta, 1992).

Disposal Strategies

The possible options for disposal of drainage water are as follows:

Disposal of Drainage Water into River System

A substantial amount of saline drainage water can be discharged into river systems having sufficient headwater flow to dilute the salts present in the effluent. The disposal in this way may in most cases be restricted to monsoon season only as most of the flow is diverted for irrigation during the remaining part of the year. Moreover, the river being the main source of drinking water and industrial needs of downstream urban areas, strict adherence of safe standards of the quality of water is necessary. The disposal of saline drainage water into the river in the non- monsoon seasons of the year may create serious environmental problems threatening the biological life in the river water. Since maximum discharge of the drainage system is expected during monsoon, the problem of safe disposal of the drainage effluent could be taken care of to a great extent.

Collection and Storage in Evaporation Tanks

Evaporation tanks are the only structures which could help dispose off the saline drainage water when no outlet for disposal to the sea or river is available. The drainage water is collected in evaporation tanks large enough to receive all the drain water from the area and allowed to evaporate. A good evaporation tank should allow a larger fraction of water to evaporate with a minimum possible seepage. In the arid and semi-arid regions about 5 per cent of the land area under reclamation need to be earmarked for evaporation tanks (Rao and Rao, 1990). Though it is an accepted practice for drainage water disposal in many countries, in the Indian context, huge loss of land and its cost of construction are the two major limiting factors in implementing this technique. Moreover, with fear of accumulation of salts and trace elements to toxic levels renders the tank technology at best a temporary solution for the drainage water disposal.

Disposal into Deep Aquifers

If geo-hydrological conditions permit, there is a distinct possibility of disposal of drainage waters in deep aquifers. Decisions to dispose off the water in this manner should, however, follow thorough investigations.

Disposal to the Sea

A drainage water carrier canal appears to be the ultimate solution to the problem of disposal. Although, presently it may appear too unrealistic, yet studies are already underway to explore the possibility of this technique in facing the future challenges of the saline water disposal problems.

MAINTENANCE

Maintenance of Open Drains

Poor design, improper construction and lack of adequate maintenance could be responsible for the failure of open ditch system. If improperly maintained even a well designed and installed system could also fail to function well. After initial construction, open drains may loose their effectiveness due to one or more of the following reasons.

— Sedimentation

— Vegetation growth

— Bank erosion

— Damage to side slopes

Maintenance has been divided into two phases i.e. preventive maintenance before failure and corrective maintenance after partial or complete failure. Preventive maintenance is usually cheaper and prevents or delays the failure of the system. Likewise timely maintenance is more economical than delayed maintenance.

Preventive Maintenance

Prevention of sedimentation, control of excessive vegetative growth in the channels, prevention of bank erosion and repair of damaged portion of the drain are some of the works that are needed to prevent drain failure. Sedimentation of drains may occur due to the sediments carried into the drains alongwith surface run-off. Minimizing the erosion in the watershed through the construction of simple drop structures, sedimentation basins and diversion ditches reduce the sediment deposits in the drains. Scouring and deposition of silt may also occur along the bed of the drain particularly at points where the grade changes sharply and there is change in flow velocities. Construction of settling basins and low check dams may be tried for such situations. These steps would reduce the clean out works in the drains.

Weed growth reduces the capacity of the drains and thus the drains fail to carry the designed discharge. Excessive weed growth in channels has been found to reduce the capacity of the channels by 75 percent or more (Fig. 4.7). The control of excessive tree growth, brush, grass and weed growth along the sides of the channel as well as weed growth in the channels is an important preventive maintenance measure.

Manual removal of weeds is not only cumbersome but is also expensive. Therefore, mechanical and chemical means are often employed to control weed growth. Mowing, grubbing, clearing, grazing, burning and spraying are some such measures by which weed growth is controlled. Burning of weeds or spraying of weedicides may be required more than once during the growing season. Since grazing is detrimental to the side slopes, this practice may be used with caution.

In a properly designed and constructed drain, the damage due to bank erosion and damage of side slopes will be minimum. Wherever necessary, seeding of the spoil banks may be

Figure 4.7 A channel completely filled with weeds. Note the trees along the banks which also cause diffculty in maintenance

undertaken to grow permanent vegetation on the spoil banks. Cattle and human trespass should be minimized to maintain side slopes.

Corrective Maintenance

After the first construction, changes in cross section, grade or alignment of the drain may be required for drain to function properly. Due to variation in soil type, side slopes at some places may cave and fill the drain. The side slope at these vulnerable points need to be reshaped to a more stable slope. Sharp curves causing meandering of the drains may necessitate modifications in the ditch alignment. Either the drain may be straightened or the curve may be flattened.

Maintenance of Subsurface Drains

An efficient and a well maintained subsurface drainage system will effectively control the water table. Regular maintenance is absolutely necessary, if the system is to perform the functions for which it is designed and installed. A buried drain very soon goes 'out of sight and out of mind'. Hence regular inspection and cleaning schedule must be followed. As compared to a open ditch system, pipe drainage system needs very little maintenance, if it has been properly installed. Because pipe drains are costly to install, these must work efficiently for a long time in order to get proper returns on the investment. Therefore, proper post-installation maintenance is of utmost importance.

Newly constructed system requires close observation during the initial one to two years of operation. After the trench has stabilized and most subsidence has taken place only periodic inspection is required. A pipe drain requiring attention can be identified during periodic inspection by the presence of seepage or soft areas over the drain, sink holes, manholes full of fine material, water backing up in the manholes, shrubs or weeds growing over the drain, iron sludge in the manholes or at the outlet etc.

Ochre Clogging and Maintenance Requirement

Clogging of drains can occur through chemical deposits. These chemicals are mainly the iron compounds. The clogging problem of this nature is generally referred to as ochre clogging. The problem can either be of the soil or of ground water origin. If the problem is of soil origin, the fear of ochre clogging recedes within 5 years. In case ochre clogging is due to ground water, it might continue for a longer period.

The problem of ochre clogging can be assessed before hand by the analysis of content of Fe in the ground water and its reaction as determined by the pH. For the same degree of problem of ochre clogging, iron content increases with increase in pH (Table 4.13).

Where ochre clogging is anticipated, it is useful to select the drainage material (pipe and the envelope) very carefully. Generally a clay tile should be preferred over a PVC pipe. Organic filter materials (say wood) containing tannin are superior to other materials. A useful proposition to minimize ochre formation is to apply lime to the back-fill material. Though it might result in loss of soil structure (reduced hydraulic conductivity) due to Fe deposits in the soil pores, the problem of clogging could be significantly reduced as a result of this operation. Self cleaning of drains with ochre clogging is difficult even for grades upto 5 per cent and therefore, drains must be cleaned periodically. It can be done either manually or by means of flushing. Chemical methods of iron removal could also be employed. In the chemical method, sulfurdioxide gas is applied at the upstream end. After the chemical reaches the downstream end, the two ends are closed for 24 hr. When the end plugs are removed, dissolved iron will come out along with the drainage water. In this case, safe disposal of the drainage water is essential as it is toxic to animals particularly to fishes.

Table 4.13 Effect of pH and Fe content of ground water on ochre formation

Fe^{++} content (mg/L)		Problem of ochre clogging
Acid zone pH <7	Alkaline zone pH >7	
< 0.5	< 1.0	Negligible
0.5-1.0	1.0 - 2.5	Slight
1.0 - 2.5	2.5 - 5.0	Medium
2.5 - 5.0	5.0 - 7.5	Pronounced
> 5.0	> 7.5	Very pronounced

REFERENCES

Dumm, L.D., 1954. Drain spacing formula, new formula for determining depth and spacing of subsurface drains in irrigated lands. **Agricultural Engineering,** ASAE 35: 726-730.

Dumm, L.D., 1964. Transient flow concept in subsurface drainage, its validity and use. **Trans. ASAE** 7: 142-146, 151.

Finkel, H.J.1983. **Handbook of Irrigation Technology**. Vol.II, CRC Press. Florida, USA. 223 p.

Gupta, I.C. 1979. **Use of Saline Water in Agriculture in Arid and Semi-arid Zones of India.** Oxford & IBH Publishing Co., New Delhi. 210 p.

Gupta, I.C. 1990. **Use of Saline water in Agriculture- A study of Arid and Semi-arid zones of India.** Revised Edition. Oxford & IBH Publishing Co., New Delhi, 308 p.

Gupta, S.K., Tejwani, K.G. and Babu, R. 1971. Drainage coefficients for surface drainage of agricultural lands for different parts of the country. **J. Irrig. Pwr.** 28: 53-60.

Hooghoudt, S.B. 1940. Bijdragen tot dekennis van enige naturrk undige grootheden van de ground. **Versi Land bouwk Onderz.** 7:449 p.

Mishra, K.L., Parwal, A.K. and Gupta, S.K. 1992. Assessment of crop tolerance to saline irrigation waters. **Bhu Jal News** 5(4) : 35-40.

Minhas, P.S. and Gupta, Raj K. 1992. **Quality of Irrigation Water-Assessment and Management**. Indian Council of Agricultural Research, New Delhi. 123 p.

Rao G.G.S.N. and Rao K.V.G.K. 1990. Disposal of subsurface drainage waters through evaporation ponds - a case study for Haryana state. **Indian J. agric. Sci**. 60: 97-100.

U S Bureau of Reclamation, SCS. 1971. Soil Conservation Service. **National Engineering Handbook, Section 16**: Drainage of Agricultural Lands, USDA, Washington, D.C.

5

LEACHING OF SALINE SOILS

In the scheme of reclamation of saline soils, subsurface drainage is an operation by which it is possible to control the water table. But it is not an end in itself. The high concentration of the salts in the root zone must be brought to an acceptable level before the land could be put under the plough. In this chapter, process of leaching is discussed and ways and means are described which help improve the leaching efficiency.

LEACHING : DEFINITION AND OBJECTIVES

Leaching of saline soils is essentially a process whereby water of low concentration is applied to displace the soil solution of relatively high concentration. The application of the excess water to pass through the root zone with the aim of pushing the salts below the root zone is defined as leaching. The leaching of salts is carried out to meet one of the following objectives:

i. To remove or to reduce the salts in the root zone commonly known as one time leaching for reclamation.

ii. To maintain the salt balance in the reclaimed or irrigated lands so that crops do not suffer due to excess salts at any time in future.

HISTORICAL BACKGROUND

The importance of leaching to reclaim salt affected lands had been recognized long ago. Mild alkali soils were reclaimed by flooding them with water for several months followed by rice transplanting (Leather, 1914; Mehta, 1940). Badly affected salt infested lands could be reclaimed by impounding water and draining away the salts. Leaching alone of alkali soils, however, failed to produce any response on crop yield or soil conditions (Singh and Nijhawan, 1932). In the Southern states, leaching alone with sugarcane crop was not successful (Talati, 1941) while leaching with drainage and green manuring with *dhaincha* (Sesbania) had great ameliorating effect (Talati, 1947). Hoon (1949) reported that *kharlands* (coastal saline soils) could be reclaimed if protected from tidal inundations and leached by rain water or through artificial irrigations. Kanwar (1962) reported results of statistically designed leaching trials to reclaim saline soils by leaching alone or in combination with gypsum and fertilizers. Leaching alone was of little benefit. Dutch scientists concluded that continuous ponding did not yield quick desalinisation in *Bhal* tract of Gujarat because of reduced infiltration rates (NEDECO, 1968). Short inundation cycles were better because infiltration rate remained high throughout. Tilling in between the inundations was beneficial.

FLUSHING FOR SALT REMOVAL

Flushing method of salt removal differs from leaching in many ways. In the former, salts are removed by surface washing of the soil as compared to pushing of soluble salts below the root zone in the latter. In case of flushing, salts move with excess rainfall or irrigation water as the water moves down the sloping land surface. The water with washed away salts is drained through a surface drain provided at the tail end. Reeve *et al.* (1955) reported that salts removed by flushing in a 100 m border strip is only 1 per cent of the total salt content of the 60 cm deep soil profile. Yadav and Agrawal (1959) observed that the volume of leachate and the quantity of sodium salt removed from a saline-alkali soil increased when the pots were given flushing in between two irrigations. Gupta (1985) observed that except for the first storm or storms following long dry spells, no appreciable amount of salt is present in the run- off occurring due to heavy rain over saline soils. The flushing method is useful only in cases where the soil permeability is low and in soils where impermeable layer exists near to the soil surface. Elgabaly (1971) recommended that in areas where there exists a salt crust of highly hygroscopic salts which may prevent use of levelling or other heavy equipment because of low bearing capacity, it is useful to flush the salts. It will permit better drying of the surface soil. A combination of flushing and leaching has been successfully used for the reclamation of lands in several projects of *Kharland* Development Board, Gujarat.

Technically, in the case of flushing, sufficient gradient in the downward direction is desirable for water to move faster. It may pose difficulties in the design of the irrigation system and for uniform spread of the water during leaching. Moreover, provision of a well designed surface drainage system is a must for the efficient working of this technique. Thus, in land locked fields, this problem is often insurmountable. Due to high water requirement, inefficient salt removal and several technical constraints; use of this technique as a means of reclamation is limited.

LEACHING FOR RECLAMATION

The only reasonable way to accomplish reclamation seems to be by applying adequate quantities of water to the soil surface, such that the excess salts present in the root zone are dissolved and transported below the root zone with percolating water. For efficient leaching, following steps should form the part of a leaching programme.

Field Preparation

Natural fields are undulated and unbunded. To enforce effective leaching, it is essential to level and bund the fields. Inadequate levelling leads to spotted desalinisation and may cause delay or failure of the reclamation. To enhance uniformity of leaching, smaller plots with temporary bunds could be constructed such that the plots could later on be adjusted within the overall layout plan of the farm. Tillage of land before and during leaching has been found beneficial because it reduces the time of reclamation as well as improves leaching efficiency.

Time of Leaching

Whether or not, sufficient water is available at a particular time determines the appropriate time of leaching. In case there is no constraint of water, the best time for efficient leaching is when the water table is deepest and evaporation is minimum. Leaching of sodium sulphate may, however, be retarded during winter season when evaporation is minimum. If water table is at shallow depths during winter, it is better to choose a period when water table is at least one m deep. In case leaching is to be accomplished through rainfall, it is natural that one has to wait for a period when excess rainfall is expected. For example, in the Indian context, at most places in arid and semi-arid zones, excess rainfall is only experienced during the monsoon season. Therefore, July- August period appears to be the best time for leaching. If, some water is available towards the beginning of this period, it could be used for leaching as it would enhance the efficiency of leaching through rainfall.

Depth of Reclamation

The depth of reclamation depends largely upon the potential land use. It could be from 60 cm for shallow rooted crops or pastures to more than 2 m for trees. For diversified cropping, the depth of reclamation usually varies from 1 to 1.5 m. It may be noted that these are the minimum recommended depths and appropriate measures against resalinisation should be adopted.

How Much to Leach

The salt status of the soil following leaching is normally determined by the tolerance of the crop to be grown soon after the reclamation. The level of salts should be such that for the planned crop and water management strategies, salt constituents of concern (salinity, specific ion concentration or alkali) should not exceed the permissible limit. If the level achieved after leaching does not permit cultivation of the proposed crop, either a more tolerant crop should be selected or additional leaching water must be applied. Most comprehensive data on salinity tolerance of crops have been published and often this data form the guidelines for crop selection in reclaimed lands (Maas and Hoffman, 1977; Maas, 1980). Gupta and Yadav (1986) worked out for the first time in India, saline water tolerance of same selected crops, the data for which had been available over a considerable period of time, irrigated under natural field conditions on different types of soils at different places widely differing in agro-climatic conditions. Crop tolerance table provides to some extent a similar information on some crops for some selected stations in India (Table 5.1). The threshold EC i.e. the EC_e up to which there would be no decline in yield , slope which represents the decline in yield per unit increase in soil salinity beyond the threshold are reported.

Table 5.1 Crop tolerance to soil salinity for working out leaching requirement

Sites	Soil type	Crop	Threshold	Slope(%)
Sampla	Sandy loam	Wheat	4.0	29.0
		Mustard	6.0	75.0
		Barley	7.0	19.0
Karnal	Sandy loam	Green gram	1.8	20.7
		Mustard	3.8	6.9
		Sorghum	2.2	10.6
Agra	Sandy loam	Wheat	8.2	19.8
		Mustard	6.1	20.7
		Berseem	3.5	12.5
		Potato	4.4	16.1
		Tomato	1.3	6.5
Dharwar	Black clay	Wheat	2.3	20.5
		Safflower	2.8	20.7
		Sorghum (W)	2.1	3.9
		Italian millet	6.1	50.0
		Sataria	1.12	2.19
Indore	Black clay	Berseem	2.00	11.22
		Safflower	2.80	5.00
		Maize	0.50	7.90
Chakuly	Clay	Rice	0.60	2.70

It is known that outcome of the entire season depends upon the initial crop stand. Therefore, most favourable conditions should be created during the germination and initial establishment stages. In case of one time leaching for reclamation, amount of salts to be drained is usually expressed as a fraction (Lf) of the amount of salts present in the profile at the start of leaching. Mathematically,

$$Lf = 1 - \frac{\text{Threshold salinity of crop}}{\text{Initial salinity of the the soil}} \tag{5.1}$$

When some yield reduction is permissible, then threshold salinity in eq.(5.1) could be replaced by the salinity at which desired yield reduction (EC_d) would occur. It can be calculated by the data reported in Table 5.1 utilising the following relationship:

$$EC_d = \frac{\text{Desired yield reduction (\%)}}{\text{Slope (\%)}} + \text{Threshold EC} \tag{5.2}$$

In case tolerance of the crop at the germination stage is different than the mean values in Table 5.1, it is proper to use the threshold salinity levels at the germination stage.

LEACHING CURVES

In planning reclamation of saline soils, the leaching curve is an useful tool for deciding whether desalinisation is technically and economically feasible and which reclamation methods are most suitable. Leaching curves could be developed with sufficient accuracy on the basis of results from simple and rapid leaching tests provided the same method of leaching is followed for preparing the curves as is intended under field conditions. A simple, straight forward way to prepare leaching curves is by running a test similar to the one used for the evaluation of the infiltration rate. Two concentric cylinders are used for this test (Fig. 5.1). The water level in the inner cylinder is kept at a constant level and water infiltrated is measured till the pre- determined amount of water has entered the soil. The experiment is usually a replicated trial. In these experiments, 3 or 4 depths of water are passed and are taken as the treatments. Soil samples in the inner cylinder are taken before and after the completion of the test for evaluating moisture content and salt status. For salt status appraisal, the most convenient approach is to measure the bulk soil electrolyte conductivity of different layers.

Fig. 5.1 A view of the set-up for preparing leaching curves

The amount of water drained below a given depth, D_W, is found by subtracting from the total depth infiltrated, the amount of water evaporated and the amount of water needed for moistening the soil. The EC_{eq} is commonly used to make the leaching curve independent of the salinity of irrigation water and the existing evaporation and drainage conditions. It is obtained by measuring the soil salinity in the 0-5 cm layer in the treatment where largest amount of water has entered the profile. It is expected that the salinity of this layer reaches a practically constant minimum soon after the leaching commences.

The dimensionless plot of $EC - EC_{eq}/EC_o - EC_{eq}$ (EC_o is the initial soil EC_e) versus D_W/D_S (D_S is the depth of the soil) will provide the desired leaching curve. This curve could be made use of in future for the test soil and conditions under which leaching test has been conducted. Data illustrating this approach for preparing a leaching curve are shown in Fig. 5.2. It shows that water requirement for leaching a light textured soil is relatively less than a heavy textured soil. The main limitation of a leaching curve in drained lands relates to the type of flow. In constructing leaching curves, flow is vertical and therefore, will be applicable to soil depths at which water movement is vertical or practically vertical. Thus, the leaching curves prepared with this method may not be applicable to strips near the drains or 1 m below the soil surface in widely spaced laterals with lateral drain depths of 1.5 m or more.

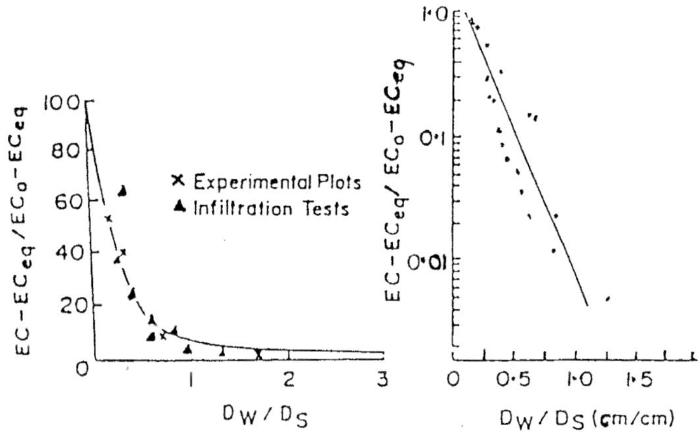

Figure 5.2 Leaching curves (a) heavy textured soil (b) medium textured soil

LEACHING AMENDED ALKALI SOILS

While saline soils could be reclaimed through leaching alone, alkali soils require application of an amendment to replace Na ions by a more favourable ion, the Ca on the exchange complex. Therefore, leaching in amended alkali soils is not only required to transport the applied chemical constituent of the amendment to the site of exchange, but also to transport the exchanged ion and soluble salts below the root zone. Commonly, gypsum and pyrites are applied as amendments to reclaim alkali soils. Leaching in the two cases would differ and therefore, each case is discussed separately.

Leaching Gypsum Amended Soils

It is believed that gypsum is sparingly soluble in water and therefore, would require huge quantities of water for leaching. Based on the solubility of gypsum in pure water, it was estimated that 150 cm of water must be passed through the profile to dissolve 19 t/ha of gypsum commonly applied in highly degraded alkali soils. Recent trials, however, do not support this contention. It has been reported that gypsum solubility in the soils increases several fold because of the presence of a sink for the calcium ions. It is estimated that to dissolve 99 per cent of gypsum in the size range of 0.5- 2.0 mm in top 30 cm of the profile would require only 36 cm of water. More recent studies indicate that water requirement for leaching gypsum amended soils may not be more than what is necessary for removing soluble salts. A good field practice for leaching gypsum amended soils is to continuously pond the fields after gypsum application for about a week and transplant rice thereafter. Remaining requirement of water for leaching would be met through deep percolation losses from the rice fields.

Leaching Pyrites Amended Soils

Pyrites is applied to the soil when the moisture content is slightly higher. Once the amendment has been applied, fields are left as such, for about a week and no leaching is carried out during this period. It allows time for the pyrites to oxidize. After oxidation is completed, water is ponded on the soil surface for about a week either in the continuous ponded mode or in intermittent ponding mode. Total time required for leaching would be slightly higher in the latter case. The leaching in this manner should be followed by rice transplanting similar to gypsum amended soils which will take care of the remaining water requirement. In view of this, it is advisable to start reclamation programme slightly early in case pyrites is used compared to gypsum.

AMOUNT OF WATER REQUIRED FOR LEACHING

In an ideal soil system without bye pass, dissolution of precipitated salts, salt diffusion constraints or hydrodynamic dispersion, the salt concentration of the soil water passing a given depth in the soil profile should drop to the concentration of the applied water when one pore volume (1 PV) of the applied water equals one pore space or one pore volume of the soil volume to be leached. The actual amount of water required under practical leaching will, however, depend mainly upon following factors :

— Amount of salt initially present in the profile

— Desired level of salt in the root zone

— Soil texture

— Type of salts

— Soil depth to be reclaimed

— Efficiency of the drainage system

— Method of leaching

As a thumb rule, a unit depth of water will remove about 80 per cent of the salts from a unit depth of the soil (Reeve *et al.* 1955). Field data, however, show a wide range depending upon the soil texture and other factors listed at the beginning of this section (Table 5.2). The amount of water required to reclaim 1 cm of the soil under field conditions varied from a low value of 0.3 cm to a high value of 2.58 cm with continuous leaching (Table 5.2). Very high values reported by Beyce (1972) appear to be a result of greater bye pass due to the cracking nature of the peat soil.

Table 5.2 Amount of water required for leaching as affected by soil type, ion monitored and method of leaching

Soil type	Method of leaching	Constituent monitored	Amount of water applied for leaching (cm/cm)	Salt removed (%)	Source
Deep Peat	Continuous	EC	2.58	80	Beyce 1972
	Intermittent	EC	4.43, 4.17	80	
	Intermittent (Sprinkling)	EC	2.25	80	
	Continuous	B	3.69	80	
	Intermittent	B	6.33	80	
	Intermittent (Sprinkling)	B	3.68	80	
Silt Loam	Intermittent	EC	0.30	*	Carter and Robbins1972
Loamy sand to Loam	Continuous	EC	0.40		Chawla (1985)

Table 5.2...

Sandy Loam (0-50)	Continuous	EC	0.30		Dahiya *et al.* (1981)
Loamy Sand (0-100)	Intermittent	EC	0.16		
Loam-silty loam	Continuous	EC	0.40		Dieleman (1963)
Clay loam silty loam	Continuous	EC	1.15		
Silty loam	Continuous	EC	0.90		
Keerange clay	Continuous	Cl (Nacl)	0.55	80	Garland (1966)
Sandy loam	Continuous	EC	0.50 (1.0 PV)	74	Gupta and Ramadurgaiaha (1984)
Peat	Continuous	EC	2.25	80	Hoffman(1980)
Clay loam	Continuous	EC	1.50	80	
Sandy loam	Continuous	EC	0.50	80	
Silty clay loam	Intermittent	EC	0.50	80	
Sandy loam	Intermittent	EC	0.50	80	
*	Intermittent	B	3.0	80	
*	Intermittent (sprinkling)	B	4.4	80	
Greenfield loam	Continuous	–	0.30 (1.5 PV)	80	Jury *et al.* (1979)
	Intermittent	–	0.68 (1.5 PV)	80	
Sandy loam	Continuous	EC	0.40	80	Khosla *et al.* (1977)
Sandy loam	Continuous	EC	0.37	80	Lefellar & Sharma, 1977
	Intermittent	EC	0.37	80	
Silty clay or clay	Intermittent	% salt	0.50	80	NEDECO (1968)
Holvilee silty clay	Continuous	*	1.2	70	Oster *et al.* (1972)
	Intermittent		0.5	70	
	Intermittent (Sprinkling)	*	0.8	80	

Contd table 5.2....

Saline organic	Continuous	EC	2.25	80	Prichard *et al.* (1985)
	Intermittent (sprinkling)	EC	2.25	80	
Silty clay loam	Continuous	EC	1.50	80	Reeve *et al.* (1955)
Yandera loam	Continuous	EC	0.70	80	Talsma (1966)
Sandy loam	Saturated	EC	0.37	80	Thokal (1990)
	Transient (Ponded)	EC	0.69	80	
	Transient (Controlled application)	EC	0.47-0.52	80	
	Continuous	EC	0.60	80	Verma and Gupta (1989)
	Intermittent	EC	0.60	80	

PV= Pore volume, * not reported

CONTINUOUS AND INTERMITTENT LEACHING

Leaching efficiency could be improved by creating unsaturated conditions during leaching and by reducing the average rate of water movement. This can be achieved through intermittent ponding of water on the soil surface. Intermittent leaching minimises loss of water through large pores and provides greater opportunities for diffusion of salts from immobile or less mobile pockets to the mobile zone. Intermittent leaching has been found to be useful in field trials and it is reported that as much as 30 to 70 per cent water required for leaching in continuous ponding could be saved. Nielsen *et al.* (1966) reported that intermittent ponding may not be useful in cases where surface layer is more restrictive than the lower layers. In such cases, even during continuous ponding, unsaturated conditions will prevail in the profile. Intermittent ponding has not been found superior to continuous ponding in alkali soils (Abrol and Bhumbla, 1973), sandy loam soil (Lefellar and Sharma, 1977) and for reclamation of saline soils formed as a result of 2.5 year of application of saline water (Jury *et al.*, 1979). Similar results were reported for leaching of boron (Beyce, 1972). Dahiya *et al.* (1981), however, reported that nearly 50 per cent saving in water could be made with intermittent leaching compared to continuous leaching in a sandy loam soil. Elgabaly (1971) reported that although intermittent ponding saved water, it took nearly 40 per cent more time for leaching compared to continuous leaching. Surface evaporation of water occurring between leaching may sometime overshadow the increased leaching efficiency brought about by intermittent leaching. Synthesis of present day data reveals that intermittent leaching may be useful in soils with large bye pass (with cracks or structured soils), poor soil permeability, deep ground water table and during periods of low evaporative demands. On the other hand continuous ponding may give reasonably good results in soils with good permeability, unstructured soils, areas with shallow saline ground water and during periods of high evaporative demand.

SPRINKLER VERSUS CONTINUOUS LEACHING

Another possibility of enhancing leaching efficiency in field soils is through the control of application rate. Sprinkler irrigation system for this purpose could be a very handy tool for applying water at pre-determined rates. Oster *et al.* (1972) reported that intermittent ponding required 1/2 the amount of water that is required in continuous ponding while intermittent sprinkling required water that was intermediate of the two. Reassessment of the data from Hoffman (1980) also indicated that under similar conditions, sprinkler irrigation required more water than intermittent ponding at least when more than 60 per cent salts are to be removed. One of the reasons for high water requirement in the case of leaching with sprinkler irrigation could be that more water is lost in evaporation as it takes more time to pass the same amount of water when water is applied at slower rate (Thokal, 1990). More water was required to leach the same amount of salt when water was applied at 1 cm/day compared to when it was applied at 2 or 3 cm/day (Fig. 5.3). In a recent experiment Prichard *et al.* (1985) did not find any superiority of leaching with intermittent sprinkling over continuous ponding. In both cases 850 mm of water was required to remove 70 per cent of salts from 100 cm of the soil profile. This could be due to high moisture content prevailing in the sprinkler treatment and to low infiltration rate (3.5 mm/day) in the continuous ponding treatment after two weeks of ponding. In the Indian context, it appears that leaching by sprinkling may not play an important role unless sprinklers are purchased and used for irrigation purposes. Moreover, data to-date suggest that similar conditions could be created through intermittent leaching. Therefore, usefulness of investment in sprinklers for leaching may be questionable.

SPATIAL VARIABILITY OF SOIL SALINITY AND ADEQUACY OF LEACHING

One of the factors that affects the amount of water required for one time leaching is the initial salinity of the soil. Because of spatial variability in the soil physico-chemical properties, it is often difficult to assign a value to the initial salinity. The argument that mean value could be an appropriate measure of the initial salinity for this purpose is doubtful. It is because in that event only 50 per cent of the land will be reclaimed, as 50 per cent of the land area is expected to have more initial salinity than the mean. In a recent study on soil salinity appraisal, Thokal (1990) studied soil salinity variation in a saline land. Commonly used flood frequency analysis procedures were used for describing salinity distribution in a 0.25 ha area. The data revealed that normal distribution and cube root transformation failed to describe the data series. Based on this analysis, it was observed that for the site under investigation, EC_2 values at 50, 80 and 90 per cent confidence level were 6.6, 9.7 and 11.6 dS/m. Utilizing the leaching curve for this site, calculated water requirement to bring the soil salinity at or below 1.0 dS/m were estimated to be 0.44, 0.52 and 0.56 cm/cm of the soil profile to be reclaimed (Fig. 5.4). It means, to reclaim 80 per cent and 90 per cent area, the increase in water requirement would be 18.2 and 27.3 per cent, respectively over and above the water needed to reclaim 50 per cent, of the area. It is therefore, opined that initially 80 per cent of the area should be reclaimed as reclamation of 50 per cent area is too less. Moreover, with 18 per cent more water, 30 per cent additional area would be reclaimed. Beyond this, amount of water required to reclaim additional area increases steeply. For example, with additional 7.7 per cent water, increase in area reclaimed is only 10 per cent i.e. 90 per cent area is reclaimed compared to 80 per cent previously. The following guidelines in this respect would be useful.

— Take soil sample in adequate number from the area to be reclaimed.

— Fit a frequency distribution to the data series. For this purpose, log-normal, Pearson type III or one step power transformation could be tried (See Chapter 3 and Annexure I).

— Calculate the initial salinity at 80 per cent confidence level and use this estimate as the initial estimate of soil salinity.

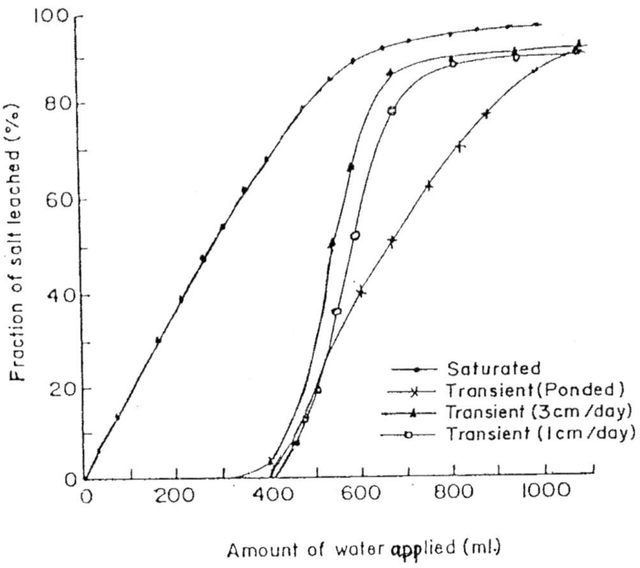

Figure 5.3 Effect of water application rates on fraction of salt leached

MODELLING LEACHING

Empirical Equations

Although mathematical models have been available for estimating water requirement for leaching, yet a number of empirical equations have been proposed based on site specific investigations and used as a practical approach for this pur-

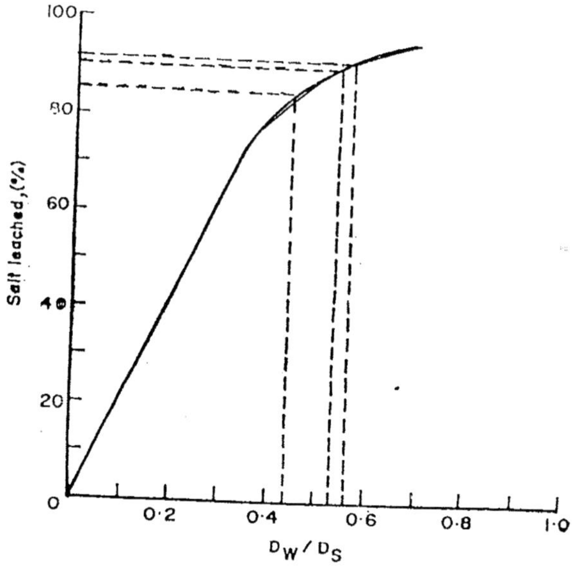

Figure 5.4 Calculation of water requirement for leaching a saline soil using a leaching curve

pose (Table 5.3). In utilizing these equations, soil parameters are not required and therefore, these equations will be applicable only to the site for which experimental data have been used in preparing the equation. Few of them, however have been generalized and extensively tested (Dieleman, 1963; Hoffman, 1980). Several of the site specific equations are reported in Table 5.3. In Table 5.3, a and b are constants and may not necessarily have the same meaning. The other parameters are d_l, d_d and d_s which stand for depth of the water applied, depth of the water drained and depth of the soil, respectively. The parameter f stands for the fraction of the salt drained. The most simple model is that of Reeve (1957) which has been commonly used by others (Dahiya et al. 1981; Lefellar and Sharma, 1977; Jury et al. 1979). A critical analysis will reveal that equations proposed by Khosla et al. (1979) and Chawla (1985) are similar to the equations based on single layer models derived on the basis of layerwise equilibrium theory with $\theta = 1/b$ and a=1. Similarly the equation proposed by Raj and Nath (1984) can also be shown to conform to the single layer model of Burns (1976). When compared with this model, value of 'a' should be 0.0 and b should be equal to 1/2. Although other equations can not be compared easily, it appears that to many field situations, single layer models could be fitted with convenience instead of trying empirical equations. Khosla et al. (1979) reported that Dieleman's and Burns models also fitted as good to his data set (r=0.92 and r=0.91) as the empirical equation (r=0.96).

Layerwise Equilibrium Theory

The approach based on layerwise equilibrium theory has been put to practice for simulating leaching of saline soils. Simple models which are presently used are probably an offshoot of the chromatography approach. In this approach, soil is divided into horizontal layers. Most are based on mass conservation in respect of both the solute and water. Basically the models consist of dividing the soil into several layers and applied water into several pulses. After each addition, the increment of water is assumed to completely mix with the soil solution in that layer. A new equilibrium is assumed to occur between the adsorbed ion or that in the solution. If the field capacity is exceeded, the excess water is assumed to move into the next layer where it again mixes with the soil water in that layer. The solution in excess of field capacity is thus assumed to move into subsequent layers until either downward percolation ceases or the required depth is reached. Each successive pulse is treated in the same manner. Simple models based on single layer and constant moisture content could be treated analytically while others require listing of a computer programme although mathematical treatment remains simple. Mathematically, all the models could be described by an equation of the form given as eq. 5.3.

$$C_{i,j} = \frac{C_{i,j-1}\,\theta_{i,j-1}\,H_i + C_{i-1,j}\,(D_j - H_{i-1}\,(\theta_{i-1,j} - \theta_{i-1,j-1}))}{\theta_{i,j-1}\,H_1 + (D_j - H_{i-1}\,(\theta_{i-1,j} - \theta_{i-1,j-1}))} \qquad (5.3)$$

It means, the salt concentration in the layer following jth application of pulse of water each of some known depth will be equal to the concentration after the j-1th pulse times the amount of water at the end of j-1 th application plus the concentration of the effluent from i-1th layer times the depth of water drained below i-1th layer divided by the total water. Here C_{ij} is the concentration of the ith layer after jth pulse, O_{ij} is the moisture content of the ith layer after jth pulse, H_i is the depth of the ith layer and D_j is the depth of water applied in jth pulse.

Table 5.3 Commonly used empirical equations for calculating leaching

Equations	Proposed & Used by	Ion
$(D_1/D_s) = a.\ \exp\ (-b(1-f))$	Beyce (1972)	EC
$(D_1/D_s) = a - b\ (1-f)$	Beyce (1972)	Boron
$(1\text{-}f) = a.\exp(-b(D_d/D_s))$	Khosla *et al.* 1979, Chawla (1985)	Chloride, EC, SAR
$(1\text{-}f)(D_d/D_s) = a$	Hoffman (1980), Jury *et al.*(1979)	EC EC
$(1\text{-}f) = a + \exp(b.D_s/D_l)$	Raj & Nath (1984)	EC
$(1\text{-}f) = a(D_s/D_l) + b$	Reeve(1957); Lefellar & Sharma (1977) ; Dahiya *et al.*(1981)	EC, ESP
$\log \dfrac{(ESP - ESP_{eq})}{(ESP_o - ESP_{eq})} = -a\ (D_d/D_s)$	Lefellar & Sharma (1977)	
$D_l = a.\log(1/(1\text{-}f))^{-b}$	Agarwal *et al.* (1982) (Reported from Russian literature)	% Salt content

D_l is depth of water drained. a and b are constants (may not be similar in different equations). f is the fraction of salts leached.

It has been shown that in practice, simple models based on the assumption that soil consists of a single layer could within practical limits of accuracy describe the process of leaching. Gupta and Pandey (1983) and Thokal (1990) discussed the limitations of the existing models and pointed out that with the assumption of a single layer, a low Peclet number is implied which is justified under field conditions because of cracks, root holes and preferential paths of flow. However, single layer models existing presently could be grouped into 4 classes depending upon the division of soil or of the water applied (Table 5.4a,b). However, in practice, similar results would be obtained from each group at least in the range of leaching required under field conditions.

Table 5.4a Grouping of the existing models on the basis of physical problem, number of pulses and number of soil layers

Group	Model proposed or referred as	Physical problem	No. of pulses	No. of layers
A	Dieleman (1963)	1*	Infinite	1
	Pandey and Gupta (1978)	2**	Infinite	1
B	Burns (1975)	1	1	Finite
	Burns (1976)	2	1	Finite
C	Modified Burns (Gupta & Gupta, 1987)	1	1	Infinite
	Modified Burns	2	1	Infinite
D	Thokal (1990)	1	Finite	1
		2	Finite	1

* uniformly distributed salts, ** surface concentrated solutes.

van Der Molen (1973) contended that complete mixing in a large soil profile is difficult and therefore, it is necessary to sub-divide the soil profile into number of layers. Accordingly, a solution for estimating concentration of the water coming out of the last layer was given as

$$Cn = Co \exp(-t/T) \sum_{i=0}^{n-1} (t/T)^{i-1} \qquad (5.4)$$

Table 5.4b Grouping of the existing models on the basis of physical problem and presentation of analytical solutions

Group	Model proposed or referred as	Physical problem	Pradiction equations
A	Dieleman (1963)	1*	1-exp (-P)
	Pandey and Gupta (1978)	2**	(P+exp(-P) -1)/P
B	Burns (1975)	1	$(1/(1+1/P.m))^{m/2}$
	Burns (1976)	2	$(1/(1+1/P.m))^{m}$
C	Modified Burns (Gupta and Gupta, 1987)	1	exp(-1/2P)
	Modified Burns (Gupta and Gupta, 1987)	2	exp(-1/P)
D	Thokal (1990)	1	$1-(1/(1+P/n))^{n}$
		2	$1-1/P+1/P(1/(1+P/n))^{n}$

* Leaching of uniformly distributed salts.
** Leaching of surface concentrated solutes.
m stands for number of sub-layers and n for number of pulses.

where Cn is the concentration of the solute coming out at the effluent end of the nth layer, t/T is the residence time in a layer and i is the dummy variable. For leaching of surface concentrated solutes (salts present in the top layer only) the solution could be written as

$$Cn = Co \exp(-t/T)\,(t/T)^{n-1} / (n-1) \qquad (5.5)$$

It may be noted that t/T is equivalent to the pore volume of the layer.

It has been shown that these models give comparable results with the hydrodynamic dispersion theory provided Peclet number in the latter is taken twice that of the number of layers in a multilayer model. It means that the number of layers increases the Peclet number or reduces the dispersive movement of salts in the profile.

LEACHING EFFICIENCY

The experience world over testifies that leaching does not take place by simple piston flow as postulated initially. If that was the case, one would need to apply only an amount equal to one pore volume to obtain complete leaching of salts present in the root zone. Several factors such as diffusion, dispersion and variation in localised flow velocities in the soil pores tend to reduce the effectiveness of leaching. Boumans and van Der Molen (1964) introduced the concept of leaching efficiency which says that part of the water used for leaching bye passes the zone of reclamation without contributing to salt removal resulting in leaching efficiencies lower than unity. It may be due to incomplete mixing or due to movement of water through the large pores (cracks, root or rat holes). The concept could be mathematically written as

$$C_d = L_e\, C_s + (1-L_e)\, C_i \qquad (5.6)$$

such that

$$C_d = C_i + (C_s - C_i)\, L_e \qquad (5.7)$$

It states that concentration of the drainage water, C_d, consists of a fraction, L_e, of the incoming water which will flow out of the root zone with the concentration C_s of the soil solution, and a fraction $(1-L_e)$ that will have its original concentration C_i. The part that has the same concentration as C_s i.e. L_e could be defined as leaching efficiency. Miscible displacement theory as verified in field and laboratory trials indicates that leaching efficiency is well below 100 per cent in relation to piston displacement. The leaching efficiency decreases with decreasing Peclet number. It also depends upon the amount of water drained. Leaching efficiency decreases as the amount of water drained increases or the reclamation proceeds. Thus, in practice leaching efficiency is a dynamic term and can not be treated as a constant (Bouwer, 1969). The assertion of Bouwer (1969) that leaching efficiency can be assigned numeric values based on soil texture appears unreasonable (Schilfgaarde, 1976; Pandey and Gupta, 1981; Gupta, 1982).

Leaching efficiency in addition to what has been stated above, also depends upon the method of water application and initial distribution of salts in the profile. Leaching efficiency would also be affected by the presence of cracks, root holes, animal burrows in the surface layer and continuous large pores in the profile as substantial amounts of water could bye pass the zone of reclamation. It appears that in the original formulation of Boumans and van Der Molen (1964), only this part has been considered towards leaching efficiency which for many practical situations may depend upon soil texture. It appears at this stage that it is necessary to distinguish between the leaching efficiency and the parameter, L_e. It is proposed that this parameter may be defined as bye pass factor or mixing efficiency while leaching efficiency could be defined as the amount of salt drained as a fraction of the amount of salt present in the profile per unit pore volume after pre- decided pore volumes of water have been drained.

RAINFALL AND SOIL SALINITY

Rainfall is of paramount importance in leaching of saline soils or soils irrigated with saline water. In the Indian context rainfall during the monsoon season substantially affects the annual salt balance. Intensity and duration of the rainfall, duration of dry spells between two rain events and soil conditions including initial salinity and its distribution in the soil profile, levelling and bunding, condition of the top soil surface and moisture content in the soil profile at the time of rainfall, all influence the process of salt leaching. In general, duration and degree of salt leaching or salt build-up over the seasons would greatly be dictated by the monsoon rains and the evapotranspiration of a particular region. Such regularities, however, could influence not only the type of salt-affected soils which develop, but also on the way in which these soils can be utilized.

The leaching due to rainfall may be substantial (Table 5.5). Kelley (1967) indicated that if the amount of precipitation is sufficient to penetrate the soil to depths below the root zone, the accumulation of salts may largely be prevented. Bhumbla *et al.* (1964) reported that in light textured soils and in regions with average annual rainfall equal to or more than 500 mm (monsoon rainfall 400 mm), highly saline waters could be used for cultivating wheat crop as the soil salinity build-up during the winter season will be washed off during the monsoon season. Leaching through rainfall is significantly more effective than leaching by continuous ponding. Gupta and Abichandani (1970), reported that after the rainfall amounting to 35 to 45 cm, the initial salt concentrations in the soils were largely reduced making 40 cm of the surface completely non-saline. It can be generalised that the depth of the soil from which the soluble salts (> 80 per cent) may be leached during the monsoon season, corresponds roughly with the total rainfall during the period.

While working on Indian data (Gupta and Abichandani, 1970), van Der Molen (1975) used the Zuur- Verhoeven procedure for predicting salinisation- desalinisation of saline water irrigated soils and concluded that nearly 30 per cent of the rainfall contributes to leaching in western Rajasthan. In this procedure leaching efficiency of the applied water has been taken as 100 per cent. Gupta and Pandey (1979) utilising the same data concluded that at 3 out of 4 stations, contribution of rainfall towards leaching was 55 per cent. The low utilization of rainfall at the fourth station (Jelwa) could be due to hard and compact nature of the soil. In these estimates single layer models were used for predicting salinisation- desalinisation. These models account for the dynamic nature of the leaching efficiency which would vary from 100 per cent initially to around 62 per cent at one pore volume.

In a similar study on leaching due to rainfall in Haryana, Gupta (1983) showed that rainfall utilisation on unbunded and unlevelled fields at Sampla and Hansi could be 10 per cent and 28 per cent, respectively. It would increase to 70 per cent and 55 per cent in case bunding and levelling of the fields is completed before the onset of the monsoon season. High run-off was responsible for the low rain water use efficiency at Sampla when fields were unbunded and unlevelled. High efficiency at Sampla observed in later stages could be due to relatively larger rainfall with relatively lesser duration and number of dry spells compared to Hansi.

Table 5.5 Change in soil salinity during monsoon season in representative soil profiles of arid and semi arid regions

Depth (cm)	EC_e (dS/m)		Depth (cm)	EC_e (dS/m)	
	Pre monsoon April/May	Post monsoon October		Pre monsoon April/May	Post monsoon October
Delhi Region [1] (Semi-arid) 1100 mm rainfall (June-September)					
Ladpur (Nangloi)			Raghopur (Najafgarh)		
0-15	11.5	1.5	0-15	32.5	3.5
15-30	9.0	0.8	15-30	6.5	3.0
30-60	8.5	9.8	30-60	7.0	2.5
60-100	7.5	1.5	60-100	7.5	3.0
100-120	7.5	2.0	100-120	7.5	3.0
Paprawat (Najafgarh)			Pahladpur (Alipur)		
0-15	11.5	0.6	0-15	12.5	0.8
15-30	7.5	0.5	15-30	7.5	0.6
30-60	8.0	0.5	30-60	6.5	0.8
60-100	8.5	0.3	60-100	5.5	0.9
100-120	8.0	0.2	100-120	5.5	1.3
Jodhpur[2] (Semi-arid) 350-450 mm rainfall (July-September)					
Dantiwada (Jodhpur)			Shikarpura (Jodhpur)		
0-20	11.0	2.6	0-20	22.2	2.1
20-40	8.4	3.8	20-40	16.5	2.4
40-60	7.2	12.3	40-60	13.4	6.6
60-100	5.9	11.3	60-90	16.0	4.3
100-150	5.0	8.3			

1. Paliwal and Yadav (1976), 2. Gupta and Abichandani (1970)

Assuming that normal cultivation practices would be followed, Minhas (1994) estimated the leaching of soils irrigated with saline water as a result of rainfall. The following three equations were fitted to the heavy, medium and coarse textured soils, respectively.

$$EC_a/EC_b = 0.1916\,(D_s/D_{rw}) + 0.0962 \quad r = 0.75 \quad (5.8)$$

$$= 0.1919\,(D_s/D_{rw}) - 0.0017 \quad r = 0.95 \quad (5.9)$$

$$= 0.1577\,(D_s/D_{rw}) + 0.0068 \quad r = 0.66 \quad (5.10)$$

In the above relations, EC_a and EC_b are the soil salinity after and before the rainy season, dS/m; D_s is the depth of the soil, m and D_{rw} is the depth of the rainfall, m. The above relations reveal that about 1.85, 0.95 and 0.76 cm of rain water would be required to leach 80% of the salt from each cm of the soil needing salt removal. On the other hand, a number of investigations have shown that for medium textured soils about 0.5-0.6 cm of water per cm of soils would be required to leach about 80% of the salts. The higher amounts of rain water are needed to compensate for the run off, rain water storage in the profile, evaporation of stored soil moisture in between two rain events and upward movement of salts during non-rainy days.

In western Rajasthan, periodical leaching has been taken advantage of by growing a salt tolerant crop during the winter season using saline ground water. Depending upon the rainfall, accumulated salts are leached during the summer season and the soil becomes fit for cultivation during the winter season. In areas where rainfall during a season is not enough to leach down the accumulated salts to the desirable extent, farmers skip the winter cropping for a year so that additional leaching could occur during the second year and soils become fit to support the cultivation of winter crops. Contribution of rainfall to soil leaching as described by Dhir (1977) indicates that fallowing is a useful technique for managing saline waters in regions with low rainfall and/or in heavy textured soils or in years with below normal rainfall.

Seasonal changes in the soil salinity for coastal saline soils revealed that in comparison to arid and semi-arid regions, leaching due to rainfall is much less in the humid regions. High run-off percentage due to heavy storms, poor rain water management pactices and shallow depths to water table contribute to lesser leaching of salts.

CONJUNCTIVE USE OF SALINE AND GOOD QUALITY WATERS

In many salt affected areas in the country, good quality water is either limited or is not available at the desired time. Semi- arid regions where large quantities of underground saline waters are available at shallow depth, offer the greatest potential for the conjunctive use of saline and good quality waters for leaching. The following conditions should be helpful for making efficient use of this technique.

i. The salinity and SAR of the saline water should be less than the salinity and SAR of the soil.

ii. Gradually lower EC and SAR water is used for leaching followed by fresh water in the final stage of leaching. Conjunctive use in this manner requires that good quality water in right quantities and at the right time should be available.

A number of studies have been conducted recently to test and quantify the saving in fresh water when conjunctive use of saline and fresh water is made compared to fresh water alone (Gupta and Pandey, 1983; Ranade, 1984; Chawla, 1985). Gupta and Pandey (1983), observed that 18-33% more water (saline+fresh) is needed in case of conjunctive use compared to when only fresh water is used. However, the requirement for fresh water is reduced by 15 - 40 percent depending upon the salinity of the ground water, initial salinity of the soil and fraction of the salts drained with saline water (Table 5.6). Chawla (1985) observed that saving in fresh water to an extent of 48.3 to 61.2 percent could be made in sandy loam soils. The total water requirement increased by 37.9 to 63.8 percent. Typical leaching behaviour of saline soils with saline, saline and fresh and with fresh water alone is shown in Fig. 5.5.

Table 5.6 A typical calculation for estimating saving in fresh water requirement

Saline soil = 100 dS/m
Leaching water = 40, 20, 0 dS/m
Final salinity desired = 4 dS/m
Fraction of the salt retained=0.04

EC of leaching water	$\frac{C_s - C_i}{C_s}$	Fraction of salt leached (%)	Leaching water* required	Fresh water required	Total water needed	Increase in water requirement (%)	Saving of water (%)
40	0.6	48	1.59	2.70	4.29	34.1	15.6
		54	2.30	2.46	2.46	48.7	23.1
20	0.8	64	1.59	2.20	2.20	18.4	31.3
		72	2.30	1.95	1.95	32.8	39.1
0	1.0	96	3.20	—	3.20	—	—

* Water requirement is in terms of pore volumes.
**C_s is the soil salinity and C_i is the salinity of leaching water.

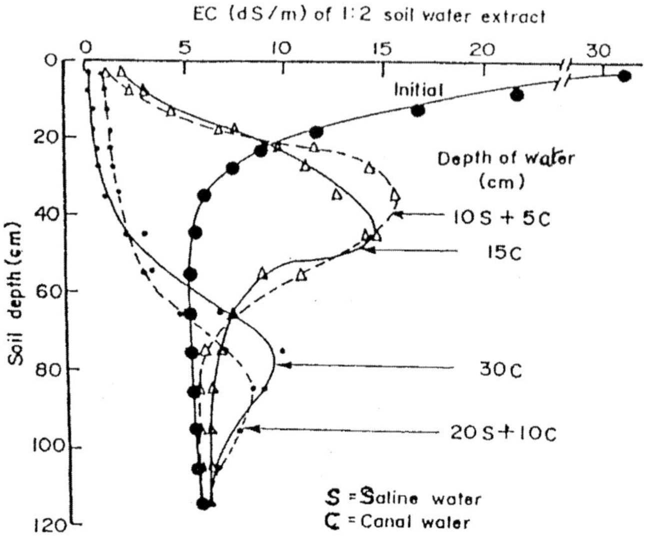

Figure 5.5 Leaching behaviour of saline soils with conjunctive use of saline and fresh water

LEACHING TILE DRAINED LANDS

Kirkham (1949) showed that for a ponded field with parallel subsurface drains, infiltration rate near the drains is much higher than near the center of the drains. It is the major cause of non-uniformity of leaching in tile drained lands. The non-uniformity is more in case drains are spaced farther apart than when these are closely spaced. Gupta (1983) observed non-uniformity of leaching in drained lands spaced 50 m apart but yields were more or less at par with distance from the drains. It was mainly because of the fact that non-uniformity occurred at or below 30 cm depth from the ground level. However, in another trial, non-uniform yields of pearl millet were observed at Sampla in the first year of cultivation in all the three drain spacings i.e. 25, 50 and 75 m. The yield was maximum upto 20 m from the drains and it declined thereafter. Laboratory tests indicated that partial ponding of the fields could to a greater extent take care of such situations and improve uniformity of leaching (Batta et al., 1980). Rao and Leeds-Harrison (1990) simulated a field situation and concluded that uniform leaching could be obtained by ponding the entire plot for 12.5 days followed by ponding the mid-half of the plot for 12.5 days and ponding of the central quarter of the plot for another 12.5 days (Fig 5.6). This scheme would result in complete and uniform leaching of the drained land. This strategy would also save nearly 50 per cent of water compared to the amount of water required to leach the field to the same level by complete ponding of the plot for the whole duration.

LEACHING FOR SALT BALANCE

Salts are added to the soil through the application of irrigation water. Sustainability of irrigated agriculture demands that amount of salts added to the soil must not be allowed to accumulate beyond a certain limit. Surplus salts must be removed by the drainage water leaving the root zone. In case mass of salts added from weathering of minerals or dissolution of salt deposits, mass of salt added as fertilizers or mass of salts lost as precipitation and mass of salt taken up by crops are neglected and it is assumed that under steady state conditions, there is no build up of salinity, then, for mass balance of salts in the root zone, we require

$$D_i C_i = D_d C_d \tag{5.11}$$

where, D_i and D_d stand for the amounts of irrigation and drainage waters and C_i and C_d are the salt concentrations of these waters. From eq.(5.11), it follows that

$$\frac{D_d}{D_i} = \frac{C_i}{C_d} \tag{5.12}$$

The leaching requirement, LR, was originally defined as the fraction of the irrigation water that must pass through the root zone to maintain soil salinity at some pre-decided level. Mathematically,

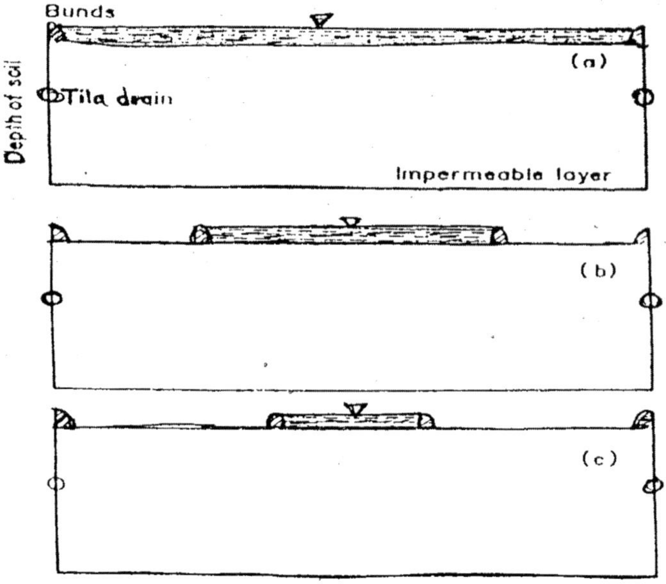

Figure 5.6 Leaching strategy for tile drained lands
(a) complete ponding (b) ponding central half of the plot and
(c) ponding central quarter of the plot.

$$LR = D_d/D_i \text{ or } C_i/C_d \quad (5.13)$$

As electrolyte conductivity (EC) of the solution is a good index of the solution concentration, solution concentration is often replaced by EC such that

$$LR = EC_i/EC_d \quad (5.14)$$

Here, EC_i and EC_d are the electrolyte conductivities of the irrigation and drainage water, respectively. Recent literature makes distinction between LR which is the theoretically average requirement for leaching and LF, which is the actual fraction of the applied water that passes through the plant root zone (Smith and Hancoc, 1986). Similarly distinction is also made between LR calculated from solution concentration as in eq.(5.13) and LR calculated from solution EC as in eq.(5.14). Smith and Hancock (1986) have shown that universal substitution of electrolyte conductivity for salt concentration is objectionable as LR estimated from C_i/C_d or EC_i/EC_d may not be equivalent.

Estimating Leaching Requirement for Salt Balance

A major requirement for estimating LR is to know C_d or EC_d. For a good approximation of these parameters for use in the LR equations, a number of suggestions have been made. Bernstein (1961) and Bower *et al.* (1968) suggested the use of average EC_e at which 50 per cent yield reduction takes place as the value of EC_d in case of forage and vegetable crops. The EC_e at which 10 per cent yield reduction occurs should be taken as EC_d in case of fruit crops. Bernstein and Francois (1973) postulated that yields are nearly insensitive to the salinity of the lower root zone and therefore, the EC_e values at 50 per cent or 10 per cent reduction in yield could be increased by 4 times before use in eq. (5.14). Schilfgaarde *et al.* (1974) discussed in detail the overestimates of, LR, by several equations notably the one proposed by Bernstein (1974) and suggested that EC_d could be assumed twice of the EC_e at which yield reduction is 100 per cent. A more logical semi-empirical model of Rhoades (1974) states that average salt concentration in the profile could be estimated from the salinity at the top and bottom of the root zone such that

$$C_S = K(C_t + C_l)/2 \qquad (5.15)$$

where, C_l is the concentration at the bottom and C_t is the concentration at the top of the root zone. Equating C_S to the average root zone salinity for a pre-determined yield reduction, LR in terms of electrolyte conductivity can be written as:

$$LR = [EC_i/(2EC_s/K - EC_i)] \qquad (5.16)$$

This equation has been used by Ayers and Westcot (1976) with the assumption that K=0.8 and EC_s=2EC_e. The proposed equation over estimated LR except at low leaching requirements. Realizing that K is a weighing factor to account for salt distribution pattern in the profile, Gupta and Yadav (1988) suggested that K in eq. (5.15) could be a function of LR and suggested that K could be related to LR as follows:

$$K = 0.7 + 0.30\ LR \qquad (5.17)$$

such that

$$EC_i/EC_s = 2\ LR/(1+LR)(0.70+0.30LR) \qquad (5.18)$$

It was shown that estimates of LR using modified Rhoades equation (Eq. 5.18) are better than the original and to a greater extent reduced over estimation at higher LR. Hoffman and van Genuchten (1983) also proposed few equations assuming that water uptake is exponential, trapezoidal or in ratio of 40:30:20:10 from the four layers of root zone. Each layer is 25 per cent of the total root zone depth. Hoffman and van Genuchten (1983) recommended that linearly averaged root zone salt concentration equation based on exponential uptake function should be used as the LR estimated with the help of this equation is close to the LF obtained under field experiments. The equation in a general form and in terms of EC, could be written as :

$$EC_s/EC_i=1/LR+(H/Z\,LR)\ln[LR+(1-LR)\exp(-Z/H)] \quad (5.19)$$

Here Z is the depth of the root zone and H is an empirical constant. The best results are obtained when H is set equal to 0.2 Z. Smith and Hancock (1988) proposed an equation essentially similar to the one proposed by Hoffman and van Genuchten (1983). It could be used to estimate the uptake weighted average root zone salt concentration. The equation is

$$EC_s/EC_i=\ln(LR)/(LR-1) \quad (5.20)$$

A comparison of four equations viz. eqs. (5.16), (5.18), (5.19) and (5.20) reveals that Hoffman and van Genuchten (1983) equation has an edge over other equations. A nomograph for calculating leaching requirement for some combinations of the soil salinity or the salinity of the irrigation water and the crop threshold is presented as Fig. (5.7a,b). This nomograph is based on eq. (5.19) and could be useful in estimating LR under most practical situations.

Appropriate Time of Leaching

The time of leaching does not appear to be critical as long as soil salinity as decided by the crop tolerance is not exceeded for extended periods or during critical growth stages. In order to decide upon the time of leaching, adequate soil and crop monitoring would be necessary. If conditions permit, annual leaching during non-cropping season should be preferred. In most cases, monsoon rainfall should be adequate to accomplish leaching. In case, excessively high saline waters are used for irrigation, leaching at each irrigation or after few irrigations or after a season may be necessary. There is considerable saving of water when periodic leaching is practised compared to regular application of extra water with each irrigation (Arar, 1971). In case of soils with low infiltration rates, it is advisable, wherever possible, to postpone leaching

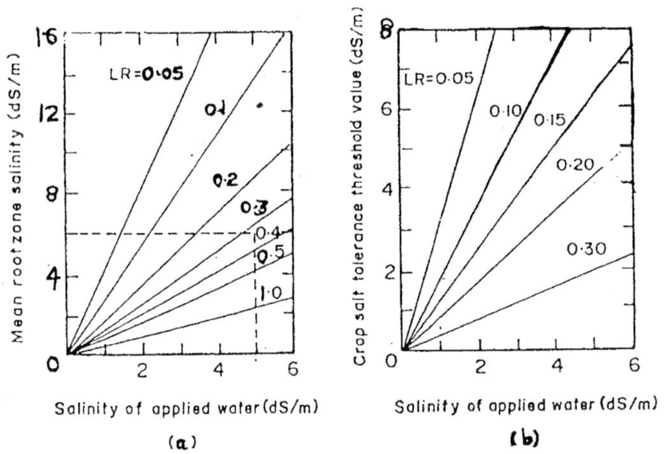

Figure 5.7. Graphical solution for estimating leaching (a) with mean root zone salinity (b) with crop salt tolerance threshold

until after cropping so as to reduce the time of water stagnation and consequently loss in crop yield.

LEACHING REQUIREMENT UNDER FIELD CONDITIONS IN INDIA

Leaching requirement concept has been developed on the basis of steady state conditions. Because of monsoon climatic conditions, steady state conditions do not get established in irrigated soils in India. Therefore, under the salinisation-desalinisation cycles experienced during a year, validity of the LR concept is doubtful.

Leaching requirement of crops usually varies from a low value of 3 per cent to 17 per cent. Hoffman and van Genuchten (1983), however, reported that LR could be as high as 29 per cent as was observed in case of alfalfa irrigated with saline water of 4 dS/m. On the other hand Bos (1977) evaluated the average water application efficiencies of five irrigation projects and reported that application efficiency is 40 per cent in one, 50 per cent in three and 70 per cent in one of the irrigation projects. Field water application efficiencies are direct indicators of the deep percolation losses under field conditions, and therefore, it is easy to visualize that a minimum LF of 0.3

is obtained under field conditions assuming uniform distribution of the irrigation water. Assuming a fair water distribution efficiency and monsoon rainfall in arid and semi-arid region, annual leaching fraction would be sufficient to meet LR requirement of various crops. With present practices in vogue, there is little chance of improving application efficiencies beyond 70 per cent under surface irrigation. It is doubtful, therefore, that there is any significance of the leaching requirement concept under the current irrigation management systems. As a testimony, field results of Jain et al.(1976) indicate that yield of wheat under irrigation with saline-sodic water (EC=4.8 dS/m, RSC=7.4 me/1, SAR=32) did not increase by application of 25 per cent excess water. Poonia et al. (1974) also observed that levels of leaching (0, 25 and 50 per cent) did not have much effect on wheat and pearl millet grown with saline waters.

In spite of the limitations, with future developments in irrigation and with increasing use of sprinkler and drip irrigation systems in the country, the LR concept may provide useful guidelines for the use of saline water. Even in surface irrigation systems, when distribution efficiency is poor, leaching fraction may play an important role in controlling salinity build-up in soils.

CROPPING DURING LEACHING

In most barren saline soils, soil salinity is too high to grow any crop. It usually takes two to six years to fully reclaim a soil due to periodic migration and accumulation of salts within the profile or from shallow saline ground water table. In order to utilize land, suitable crops should be identified and grown during the period of leaching. This state is usually referred to as the leaching-cropping stage. During this period, crops are selected on the basis of salt tolerance, tolerance to waterlogging, water requirement-water supply scenario and economic value of the crop. A very common recommendation that salt tolerant crop should be grown during initial years, should be viewed carefully before implementation. For example barley, no doubt is a salt tolerant crop but it also requires less water owing to its low evapotranspiration. Thus, there is every likelihood for the salinity level in the surface to increase compared to a crop like wheat or berseem (clover) which require relatively more water to meet their evapotranspiration requirement. Similarly, rice crop is often considered inappropriate because it is less tolerant to salts. On the other hand, if water could be arranged for a season, due to high water requirement of this crop, salt level in the profile could be largely reduced to allow a greater choice of crops in the forthcoming season following the rice crop. It is recommended, that one rice crop should be included in a long term crop rotation on such lands even after reclamation.

The cropping of the virgin barren lands could begin either with the monsoon season or from the winter season depending upon how much reclamation could be achieved before the optimum sowing period. Dahiya *et al.* (1981) transplanted 6 week old seedlings just after submergence and without any previous leaching except due to rainfall. A good rice crop was obtained in all the plots (with or without previous leaching) mainly because of the fact that soon after submergence, salinity in the top 10 cm layer reduced to 13.7 dS/m whereas 6 week old seedlings could tolerate a salinity level of 14.0 dS/m (Pearson, 1959). During the rice crop, 223 mm of water passed through 100 cm soil depth and therefore, good wheat and sesbania crops could be taken subsequently. The yield differences between intermittent and continuous ponding were non-significant mainly because the salinity in the two treatments dropped to levels well below the tolerance limits of the crops grown. On the other hand when reclamation was started with sesbania or wheat, the amount of water applied during pre-sowing leaching had a significant effect on the crop yield (Table 5.8). It may be seen that there is no significant difference if 43.5 cm or more water was applied, although numerically, increase in yield was obtained upto 63.5 cm of water depth.

Table 5.8 Effect of application of leaching water on crop yield

Leaching	Sesbania green manure (q/ha)	Leaching water (cm)	Wheat grain (q/ha)
0	33.9	33.5	1.8
10	86.3	43.5	27.4
30	113.1	63.5	33.5
50	119.5	83.5	32.1
70	117.2	103.5	35.0
CD at P = 0.05	8.62		

Source : Dahiya *et al.* (1981)

In similar trials, Gupta (1981) experimented with pearl millet and wheat crops grown in sequence. Gupta (1983) also tried reclamation, starting from the winter season. In this case, wheat was grown as the first crop. In the first case, nearly 43.2 cm of rain water was conserved in bunded and levelled fields and an additional 17.0 cm of water was applied to irrigate the pearl millet crop. The pearl millet yield was very poor when grown without pre-leaching compared to when 20.0 cm of water was applied as pre-sowing leaching. The yield improved from 0.3 t/ha to 1.25 t/ha. The overall yield was poor because initial salinity was high and drainage was not provided which leads to the resalinisation of the plots during the crop growing season. Notably, leaching carried out during the monsoon season helped better growth of wheat following pearl millet crop (Table 5.9).

For growing wheat as the first crop, plots bunded and levelled before the onset of the monsoon season were used. As much as 80 to 90 per cent of initial salts were leached during the monsoon season. However, initial sampling showed wide variation in soil salinity in various plots. To account for this variation, plots were divided into three categories i.e. plots with low, medium and high salinity. The effect of different amounts of water applied as pre-leaching treatment on wheat growth is shown in Fig. 5.8. The amount of water applied as pre-leaching treatment resulted in significant differences in yield only in case of medium and high salinity plots. Similar results were obtained in plots where barley was grown as the test crop. As the threshold soil salinity for wheat and barley is estimated at 4.5 and 6.75 dS/m respectively, significant differences were not observed in the plots with low salinity. The minimum amount of water applied could lower the soil salinity at or below the threshold level.

Table 5.9 Effect of differential water application during summer on wheat yield during winter

Water applied during monsoon (cm)		Yield (t/ha)	
Rainfall	Irrigation	Pearl millet	Wheat
32.0	0	–	1.2
43.2	0	–	3.2
43.2	17	0.30	3.8
43.2	20 + 17	1.25	4.5

Source : Gupta (1983)

PRACTICES FOR ENHANCING LEACHING EFFICIENCY

Leaching of saline soils is the only practical means for bringing the soils under the plough. To reclaim more soil with limited amount of water or to achieve early reclamation, a number of practices have been used under field conditions. The useful practices for this purpose are summarized in Table 5.10.

PITFALLS

Quite often, leaching is considered as simple application of excess amount of water which is applied before sowing of the crops. In practice, such simple interpretation of the technique has resulted in delays or even complete failure of the reclamation programmes. Practically speaking, leaching is beneficial only when applied as a package. One or more of the following pitfalls are very often identified in the design of reclamation programmes.

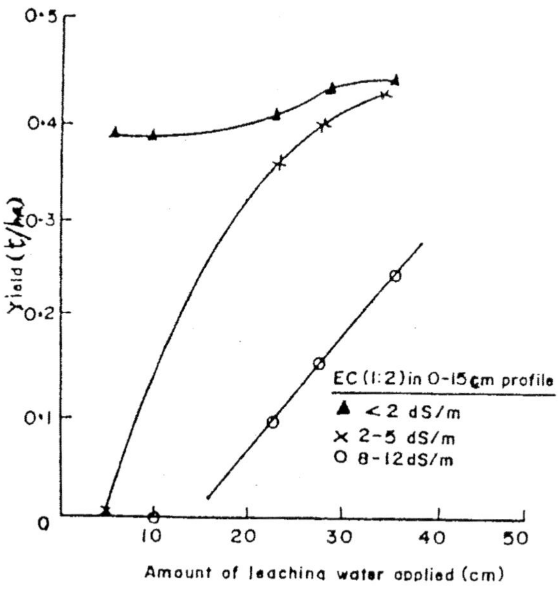

Figure 5.8 Amount of water applied for leaching and its effect on growth of wheat

i) Leaching without adequate drainage often leads to severe problem of waterlogging and soil salinisation. Right type of drainage system with appropriate capacity must be installed at the right depth before starting a reclamation programme including leaching.

ii) Inadequate bunding, levelling, field layout and slope often result in spotted leaching.

iii) Arrangements for water in right quantities and at the right time should be ensured. Inadequate application causing partial leaching may lead to failure of the whole programme.

iv) If delay is feared due to low permeability of soils, then cultural practices which enhance infiltration rate and/or suitable amendments need to be applied at the right moment. Many a times this aspect is ignored.

v) Steps are not taken to minimize secondary salinisation which in most cases is the principal mechanism leading to delay/failure of the leaching programme.

Table 5.10 Practices for modifying leaching behaviour

Practice	Effect
Subsoiling	• Improves infiltration rate
	• Removes/reduces natural channels
Tillage with inversion	• Improves infiltration rate of the soil
	• Physically moves the surface layer with high concentration of salts to lower depths
Chemical amendments	• Improve infiltration rate
	• Control surface crusting
	• Maintain soil structure
Breaking hard pan (Auger hole piercing)	• Improves water movement through the hard pan
	• Reduces chances of water table build-up
Irrigation practices	• Improve leaching efficiency by maintaining high moisture content
	• Situations resembling intermittent leaching could be created

REFERENCES

Abrol, I.P. and Bhumbla, D.R. 1973. Field studies on salt leaching in a highly saline-sodic soil. **Soil Sci.** 15:429- 433.

Agarwal, R.R., Yadav, J.S.P. and Gupta, R.N. 1982. **Saline and Alkali Soils of India.** Indian Council of Agricultural Research, New Delhi. 286 p.

Arar, A. 1971. Irrigation and drainage in relation to salinity and waterlogging. **Irrig. & Drg. Paper 7.** Salinity Seminar, Baghdad, FAO Rome, 86-111.

Ayers, R.S. and Westcot, D.W. 1976. Water Quality For Agriculture. **FAO Irrig. and Drain. Paper 29.** FAO, Rome.

Batta, R.K.; Pandey, R.N. and Gupta, S.K. 1980. Studies to improve uniformity of leaching under tile drainage system. **Annual Report,** CSSRI, Karnal, p. 83.

Bernstein, L. 1961. Tolerance of plants to salinity. **Am. Soc. Civil Engg. Proc.** 87(IR4) : 1-12.

Bernstein, L. 1964. Salt tolerance of plants. **USDA Agri. Inf. Bull. No. 283.** USDA, Riverside, California, USA.

Bernstein, L. 1974. Crop growth and salinity. *In* **Drainage for Agriculture. (Ed.J. van Schilfgaarde). Agronomy Monograph, 17.** Am. Soc. Agronomy, Washington : 39-54.

Bernstein, L. and Francois, L.E. 1973. Leaching requirement studies: sensitivity of alfalfa to salinity of irrigation and drainage waters. **Soil Sci. Soc. Am. Proc.** 37:931-43.

Beyce, O. 1972. Drainage of salty soils. **Irrigation and Drainage Paper 16;** FAO Rome, 63-67.

Bhumbla, D.R., Kanwar, J.S., Mahajan, K.K. and Singh, B. 1964. Effect of irrigation with different sodium and salinity hazards on the growth of crops and the properties of soils. **Proc. Symp. Problems of Indian Arid Zone,** Jodhpur, 174-78.

Bos, M.G. 1977. Arid land irrigation in developing countries- environmental problems and effects. (Ed. Worthington E.B.). **Pergamon Press,** Oxford, 193-204.

Boumans, J.H. and van Der Molen, W.H. 1964. Out watering- behoefts can bevolede gronden in verband met hun zouthuishouding. **Landis Tydrschn**, ft. 76 : 880-887.

Bouwer, H. 1969. Salt balance, irrigation efficiency and drainage design. **ASCE Proc.** 95 (IR1) : 121-144.

Burns, I.G. 1975. An equation to predict the leaching of surface -applied nitrate. **J. Agril. Sci. Camb.** 85:443-454.

Burns, I.G. 1976. Equations to predict the leaching of nitrate incorporated to a known depth or uniformally distributed throughout soil profile.**J. Agril. Sci. Camb.** 86:305-311.

Carter, D.L. and Robbins, C.W. 1972. Salt outflows from new and old irrigated lands. **Soil Sci. Soc. Am. J.** 42:627- 632.

Chawla, K.L. 1985. Salt and water transport under high water table conditions. **Unpublished Ph.D. thesis. Kurukshetra University, Kurukshetra.**

Dahiya, I.S., Malik, R.S. and Singh, M. 1981. Field studies on leaching behaviour of highly saline-sodic soil under two modes of water application in presence of crops. **J. Agric. Sci. Camb.** 97:383-389.

Dahiya, I.S., Singh, M. and Hajrasuliha, S. 1980. Simultaneous transport of surface-applied salts and water through unsaturated soils as affected by infiltration, redistribution and evaporation.**Soil Sci. Soc. Am. J.** 44:223.

Dhir, R.P. 1977. Saline waters, their potentiality as a source of irrigation. *In* **'Desertification and its control'** CAZRI, ICAR, New Delhi, 130-141.

Dieleman, P.J. 1963. Reclamation of salt affected soils in Iraq.**Voenman,** Wageningen. 175-190.

Elgabaly, M.M. 1971. Reclamation and management of salt affected soils. **Salinity Seminar, Baghdad. FAO Irrigation and Drainage Paper 7**:50-79.

Garland, K.R. 1966. Reclamation of waterlogged saline lands in Northern Victoria, Australia-A review of recent field experiments. **6th Congr. Irrig. and Drain.** 19: 147- 157.

Gupta, I.C. and Abichandani, C.T. 1970. Seasonal variations in the salt composition of some saline water irrigated soils of western Rajasthan. **J. Indian Soc. Soil Sci.** 18: 428-35.

Gupta, I.C. and Yadav, J.S.P. 1986. Crop tolerance to saline Irrigation waters. **J. Indian Soc. Soil Sci.** 34 : 279-286.

Gupta, S.K. 1981. Evaluation of salt and water balance in saline soils without provision of drains. **Annual Rep.** CSSRI, Karnal.112-114.

Gupta, S.K. 1982. Evaluating solute movement parameters in one dimensional flow through soils. **Unpublished Ph.D. Thesis.** J.N.T. University, Hyderabad.

Gupta S.K. 1983. Evaluation of leaching and resalinisation behaviour of saline soils in relation to crop growth. **Annual Rep.** CSSRI, 114-115.

Gupta S.K. 1985. Dynamics of salts in saline water irrigated soils. **J. Instt. Engrs.(India)** 41: 17-22.

Gupta, S.K. 1992. Grow dhaincha as a cover crop in saline soils. **Indian Farming** 42 (66) : 29.

Gupta, S.K. and Gupta, I.C. 1987. **Management of Saline Soils and Waters**. Oxford and IBH Pub. Co. Pvt. Ltd. New Delhi.

Gupta, S.K. and Ramadurgaiah, D. 1984. Mathematical model for leaching and its parameter identification under laboratory and field conditions. **J. Instt. Engrs. India** 64:68-71.

Gupta, S.K. and Pandey, R.N. 1979. A simplified equation for reactive solute transport in soil. **J. Agric. Sci. Camb.** 92:587-589.

Gupta, S.K. and Pandey, R.S. 1983. A mathematical model for evaluating leaching of saline soils when saline and fresh water are used in conjunction. **Indian J. Agric. Sci.** 53 : 701-706.

Gupta, S.K. and Yadav, R.N. 1988. Note on leaching requirements for salt balance in irrigated lands. **Curr. Agric.** 12 : 101-104.

Hoffman, G.J. 1980. Guidelines for reclamation of salt affected soils. **Proc. Inter-American Salinity and Water Mgmt. Tech. Conf.** Juarez, Mexico, 49-64.

Hoffman, G.J. and van Genuchten M.Th.1983. Soil properties and efficient water use: water management for salinity control. *In* **Limitations of Efficient Water Use in Crop Production**. Am. Soc. Agronomy, Madison, USA.

Hoon, R.C. 1949. The physico-chemical changes in deteiorated soils as the result of soil reclamation operations. **Cent. Bd. Irrig. J.** 6 : 50-3.

Jain T.C., Saxena G.S., Jain K.C. and Totawat K.L. 1976. Effect of crop rotation, farm yard manure and the quantity of saline irrigation water on the yield of wheat and physio-chemical properties of the soil. **Indian J. Agric. Res.** 10: 1-9.

Jury, W.A.; Jarrell, W.M. and Devitt, D. 1979. Reclamation of saline-sodic soil by leaching. **Soil Sci. Soc. Am. J.** 43: 1100- 1106.

Kanwar, J.S. 1962. Reclamation of saline-alkali soils. **Proc. Symp. Salinity and Alkalinity Soil Prob.** I.A.R.I., New Delhi.:54-60.

Kelley, W.P. 1967. Use of saline irrigation water. **Soil Sci.** 95 : 385-399.

Khosla, B.K.; Gupta, R.K. and Abrol, I.P. 1979. Salt leaching and the effect of gypsum application in a saline-sodic soil. **Agric. Water Mgmt.** 2: 193- 202.

Kirkham, Don. 1949. Flow of ponded water into drain tubes in soil overlying an impervious layer. **Trans. Am. Geophys. Union,** 30: 369- 385.

Lefellar, P.A. and Sharma, R.P. 1977. Leaching of a highly saline-sodic soil. **J. Hydrol.** 32: 203- 218.

Leather, J.W. 1914. Investigations on usar land in the United Provinces. **Govt. Press, Allahabad.**

Maas, E.V. 1980. Crop tolerance to saline soil and water. In: Prospectes for Biosaline Research. **Proc. US-Pak Biosaline Res. Workshop,** Karachi, Pakistan: 205- 211.

Maas, E.V. and Hoffman G.J. 1977. Crop salt tolerance: Current assessment. **J. Irrig. Drainage Div.** 103 (IR 2) :115- 134.

Mehta, M.L. 1940. The formation and reclamation of thur land in the Punjab. **Res. Publ. Punjab Irrig. Res. Inst.** 3 : 14 p.

NEDECO, 1968. Reports on Bhal reclamation scheme of state of Gujarat, India. Vol. I to IV. **Grontmij n.v. de Bilt, the Hague, Netherlands.**

Nielsen, D.R., Biggar, J.W. and Luthin, J.N. 1966. Desalinisation of soils under controlled unsaturated flow conditions. **6th Int. Congr. Irrig. and Drain.** New Delhi II: 15-24.

Oster, J.D., Willardson, L.S. and Hoffman, G.J. 1972. Sprinkling and ponding techniques for reclaiming saline soils. **Trans ASAE** 15 : 1115-1117.

Paliwal, K.V. and Yadav, B.R. 1976. Irrigation water quality and crop management in the Union Territory of Delhi. **Res.Bull. 9** IARI, New Delhi, 170 p.

Pandey, R.N. and Gupta, S.K. 1978. Equations to predict leaching of soluble salts in saline soils. **J. Agric. Sci. Camb.** 91:131- 134.

Pandey, R.N. and Gupta, S.K. 1981. Salt balance and leaching efficiency in farm irrigation systems. **J. Indian Soc. Soil Sci.** 29: 50-54.

Pearson, G.A. 1959. Factors influencing salinity of submerged soil and growth of rice. **Soil Sci.** 87:198- 206.

Poonia, S.R., Jhorar, L.R., Nath J. and Khanna, S.S. 1974. Effect of quality of irrigation water, leaching level and farm yard manure on performance of wheat and pearlmillet. **Indian J. Agric. Sci.** 44: 854-859.

Prichard T.L., Hoffman, G.J. and Oster J.D. 1985. Reclamation of saline organic soil. **Irrig. Sci.** 6:211-220.

Raj, M. and Nath, J. 1984. Estimation of water requirement for reclaiming saline soils. **J. Indian Soc. Soil Sci.** 32:249-252.

Ranade, D.H. 1984. Leaching phenomenon in sandy loam soils. **Unpublished M.Tech. Thesis,** I.I.T. Kharagpur.

Rao, K.V.G.K. and Leeds-Harrison, P.B. 1990. Water management strategies for improving leaching in subsurface drainage. *In* **Management of irrigation systems (Tyagi et al.Eds.)** Proc. Nat. Seminar Irrigation Management, Karnal: 221- 232.

Reeve, R.C. 1957. The relation of salinity to irrigation and drainage requirements. **3rd Congr. Irrig. and Drain.** 5:10.175-10.187.

Reeve, R.C., Pillsbury, A.F. and Wilcox, L.V. 1955. Reclamation of a saline and high boron soil in the Coachella Valley of California. **Hilgardia**, 24: 69-91.

Rhoades, J.D. 1974. Drainage for salinity control. *In* **Drainage for Agriculture. (Ed.J. van Schilfgaarde). Agronomy Monograph 17**. Am. Soc. of Agronomy. Wisconsin, U.S.A., 433-461.

Schilfgaarde, van. J., Bernstein, L., Rhoades, J.D. and Rawlins, S.L. 1974. Irrigation management for salt control. **J. Irrig. Drain. Div. ASCE.** 100 (IR3):321- 338.

Schilfgaarde, Van J. 1976. Water management and salinity. Prognosis of Salinity and Alkalinity. **Soils Bull. 31.** FAO Rome. 53-68.

Singh, D. and Nijhawan, S.D. 1932. A study of the physico- chemical changes accompanying the process of reclamation in alkali soils. **Indian J. Agric. Sci.** 2:1-18.

Smith, R.J. and Hancock, N.H. 1986. Leaching requirement of irrigated soils. **Agric. Wat. Manage.**, 11:13-22.

Talati, R.P. 1941. Damaged lands in the Deccan and their classification. **Indian J. Agric. Sci.** 11: 959- 977.

Talati, R.P. 1947. Field experiments on the reclamation of salt lands in Baramati of Bombay Deccan. **Indian J. Agric. Sci.** 17: 153-174.

Talsma, T. 1966. The effect of soil physical condition on reclamation of saline land. **6th Congr. Irrig. and Drain.** New Delhi 19: 83-89, 91.

Thokal, R.T. 1990. Strategies of leaching in highly saline soils. **Unpublished M. Tech. Thesis,** G.B.P.U.A.T., Pantnagar.

van Der Molen, W.H. 1956. Desalinisation of saline soils as a column process. **Soil Sci.** 81:19-27.

van Der Molen, W.H. 1973. Salt balance and leaching requirement. In Drainage Principles and Applications: Theories of Field Drainage and Watershed Runoff. **Int. Institute For Land Recl. and Improvement.** Wageningen. 11:62- 99.

van Der Molen, W.H. 1975. The Zuur-verboeven method for calculating water movement in saline soils and its application to some indian data. **J. Res. Hisar** 5 : 195-9.

Verma, S.K. and Gupta, R.K.,1989. Leaching behaviour of saline clay soil under two modes of water application. **J. Indian Soc. Soil Sci.** 37:803-808.

Yadav, J.S.P. and Agrawal, R.R. 1959. Dynamics of soil changes in the reclamation of saline alkali soils of the Indo-Gangetic alluvium. **J. Indian Soc. Soil Sci.** 7:213-222.

6

CROP PRODUCTION AND ALTERNATE MANAGEMENT PRACTICES

The foregoing chapters have discussed the diagnostic criteria, ways and means by which water table can be controlled and leaching through which excess soluble salts can be brought to a level that a given crop can be grown. Prevention and control of soil salinity in the root zone are necessary for sustainable irrigated agriculture and to save our land and water resources from degradation. It may be prudent in some cases where such steps are not feasible due to one reason or the other to resort to a method that allows to live with the salinity problem over a short or a long period of time. The management and crop production aspects for both the situations are discussed in this chapter.

FACTORS AFFECTING CROP PRODUCTION

In saline soils plant growth is adversely affected due to the following factors:

i. Excess soluble salts depress the water potential of the nutrient medium and hence restrict the water uptake by plant roots.

ii. High concentrations of specific ions can be toxic and induce physiological disorders (e.g. Na^+, Cl^-, SO_4^{--}, Mg^{++}, B).

iii. High salinity may affect different metabolic processes such as CO_2 assimilation, protein synthesis and respiration.

iv. Imbalance in the uptake of ions.

Nature of Salt Stress and Osmotic Potential

The osmotic potential of the soil solution is the primary salinity variable to which crops respond. The total soil moisture stress consists of two major components, i.e. the matric potential and the osmotic potential. Because of the additive nature of the effects of matric and osmotic potentials, the sum of the two potentials provides total soil moisture stress. Whereas, osmotic potential predominates in saline soil low-tension treatments, matric potential predominates in non-saline high-tension treatments (Fig. 6.1). Unless the specific ion or nutritional imbalance is caused by the salinity, the crop yields are influenced to equal degree under similar variations of the total soil moisture stress, be it because of soil salinity or due to delayed irrigation. It has been noticed that salinity produces symptoms which are similar to the one caused by

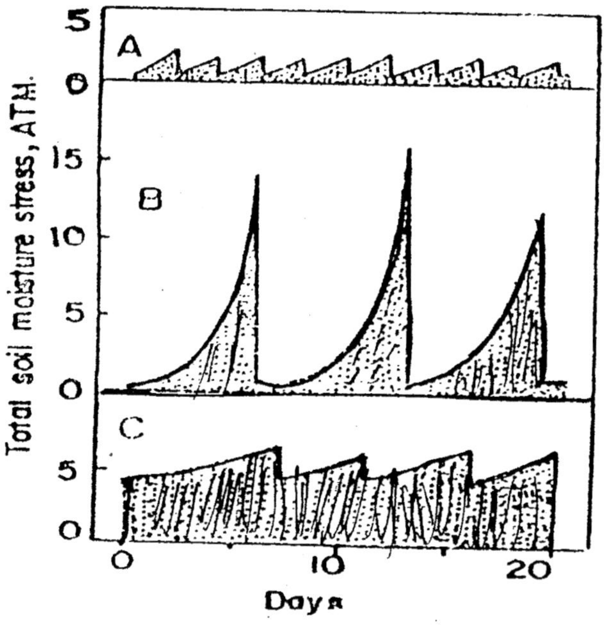

Figure 6.1 Characteristic variations of osmotic potential in irrigated soils : (A) Frequently irrigated non-saline soil, (B) Infrequently irrigated non-saline soil, and (C) saline soil

drought conditions. For example, symptoms such as stunting of plants, a lower shoot-root ratio, and smaller, darker green (bluish-green) leaves are commonly observed in saline environment which closely resemble the symptoms caused by drought conditions. In view of this, a school of thought considers that the salinity affects the plants as a result of disturbance in the plant- water balance. This situation is commonly referred to as a 'physiological drought'.

Osmotic potential of the soil solution at field capacity, is related to the electrolyte conductivity of the saturation extract, EC_e and is given by:

$$\sigma_{fc} = -7.8\ EC_e - 0.046\ EC_e^2 \qquad (6.1)$$

This relation is true at 25^0C and assumes that the water content at field capacity is equal to half of the water content at saturation. The osmotic potential is expressed in J/kg and EC_e in dS/m. A simple relationship between the osmotic pressure in atmosphere (O.P.) and electrolyte conductivity of the saturation extract in dS/m at 25^0C can be approximated by the simple relationship:

$$O.P. = 0.39\ (EC_e \times 10^3) \qquad (6.2)$$

Toxicity and Plant Tolerance

The specific ions affect the plants in numerous ways and the actual mechanisms in many cases are not well understood. The injury as a result of specific ions would depend upon the type of the salt and the plant species or even the plant varieties. Specific ions may influence respiratory pathways as in pea roots, inhibit the production of chlorophyll and carotene as in grapefruit. The most chloride sensitive plants may be injured when Cl^- in soil saturation extract exceeds 5 to 10 meq/l. The chloride tolerant plants, however, can tolerate Cl^- concentration upto 30 meq/L. Excess sodium can cause nutritional disorders by reducing uptake of calcium and magnesium. Sodium in addition to affecting plant nutrition is known to adversely affect the soil physical conditions imparting low permeability to water and air and accumulation of black solubilized organic matter on the surface. This limits water and air infiltration and root penetration. Boron may be toxic at concentrations only above 1.0 mg/l required for normal plant growth. Lithium and fluorine may be toxic to citrus and at other sensitive crops in micro-concentrations.

CRITERIA FOR RATING CROP SALT TOLERANCE

The ability of crops to tolerate salinity is usually determined by measuring one or more of the following four parameters :

— Germination

— Plant survival

— Absolute plant growth or yield

— Relative growth or yield

Large number of investigations on differential response of crops or varieties have been reported at germination stage and attempts made to extrapolate the tolerance limit for final performance. More often than not such attempts are likely to be frustrating because tolerance characteristics differ from one stage of growth to another.

Plant survival on saline soils has been widely used by ecologists to judge the tolerance of natural species growing on uncultivated lands. To the farmer, however, survival without an economic yield obviously has little value. Indirectly, this criterion is potentially useful because plant breeders can use it as the basis for selecting salt tolerant lines and cultivars. Absolute yields are certainly of greatest interest to the farmers. Absolute salt tolerance is directly related to the maximum growth or yield potential of a given plant under high salinity regardless of its growth rate under nonsaline conditions. Under salinity stress, plant growth may decrease substantially without affecting its ranking for salt tolerance.

Relative yield is the yield of a crop grown on a saline soil expressed as a fraction of that would be obtained under non saline conditions. In such evaluations, it is assumed that other factors affecting crop production do not change. This criterion permits a comparison of crops whose yields may be expressed in quite different units. The reliability of relative yield data depends upon whether yield reductions are unaffected by the level of other essential nutrients. Relative tolerance may also contribute to the overall fitness of the species to produce satisfactorily in a range of saline and non saline environments, a character referred to as environmental plasticity.

The only agronomically important plant response measurement for establishing salt tolerance of a crop is the commercial yield. For crops that have utility both in its vegetative form and in the seed yields e.g. alfalfa, pearl millet, corn etc., it is advisable to assess responses to salinity on both the accounts. Too often vegetative responses are unreliable for predicting salt effects on fruit or seed production (Maas and Hoffman, 1977). Salinity may decrease grain yields of corn and rice without appreciably affecting straw yields.

Salt Tolerance Function

According to Maas and Hoffman (1977), the relative yield of a crop could be expressed in the form of a piece-wise linear regression equation which could be described as

$$Y = 100 \qquad 0 \leq EC \leq EC_t \qquad (6.4)$$

$$Y = 100\text{-}S\,(EC\text{-}EC_t) \qquad EC_t \leq EC \leq EC_o \qquad (6.5)$$

$$Y = 0 \qquad EC \geq EC_o \qquad (6.6)$$

Here, Y is the relative yield of the crop, EC is the electrolyte conductivity of the saturation extract, EC_t is the threshold EC upto which there is no decline in the yield of the crop, S is the slope and represents the yield reduction per unit increase in the EC of the soil solution and EC_o is the EC at or beyond which crop fails to give any yield. A plot of the piecewise linear model indicating yield variation with soil salinity is shown in Fig. 6.2. Although the model is known to have some limitations in describing salinity response functions for some crops, nevertheless it has been widely tested and used. Recently, this model has been validated to describe crop response to specific ion toxicity due to boron, exchangeable sodium percentage (Gupta and Sharma, 1990) and for salinity of the irrigation water (Misra *et al.*, 1990). On this basis, crop response functions for various crops under different agro-climatic conditions have been derived and relevant parameters are presented in Table 5.1. More comprehensive data have been published by Maas and Hoffman (1977).

On this basis, salt tolerance data are given in terms of two key parameters:

1) Threshold salinity, i.e. the maximum soil salinity, that does not reduce yield below that produced under non saline conditions; and

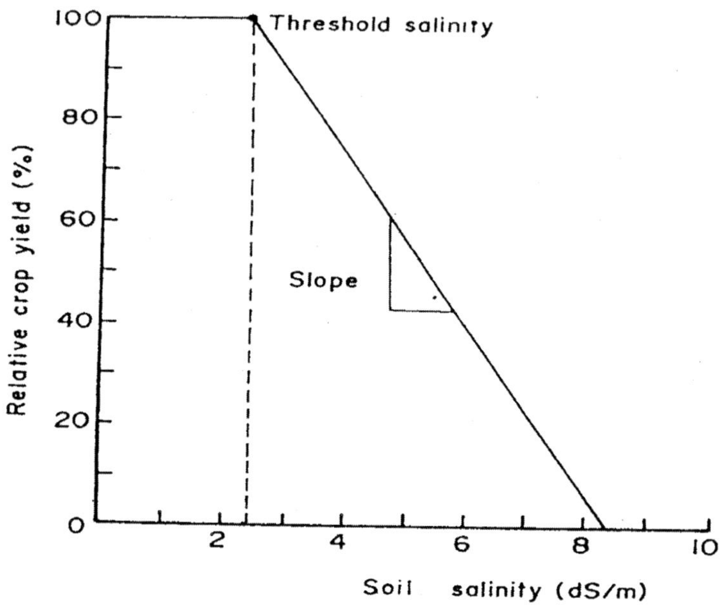

Figure 6.2 Crop response function based on piecewise linear model

2) Yield reduction per unit salinity greater than the threshold.

With these two parameters, the relative yield (Yr) at any given salinity (ECe) exceeding the threshold can be calculated with the following equation:

$$Yr = 100 - S\,(EC_e - EC_t) \qquad (6.7)$$

Any rating, however, can not provide an accurate estimate of the expected yield as the yield depends not only upon level of salinity but also upon many other cultural and environmental factors. The interactions between salinity and various soil, water, and climatic factors could modify the ability of the plant to tolerate salinity.

It should be clear that salt tolerance rating serves only as guidelines to the relative salt tolerance among crops. They can not nor they are intended to be used for estimating economic crop losses. Qualitatively crops have been grouped into 4 groups depending upon their tolerance to soil salinity (Table 6.1).

Table 6.1 Crop groups based on response to salt stress

Sensitive groups		Resistant groups	
Highly sensitive	Medium sensitive	Medium tolerant	Highly tolerant
Lentil	Radish	Spinach	Barley
Mash	Cowpea	Sugarcane	Rice (transplanted)
Chickpea	Broadbean	Indian mustard	Cotton
Beans	Vetch	Rice (direct sowing)	Sugarbeet
Peas	Cabbage	Wheat	Turnip
Carrot	Cauliflower	Pearlmillet	Tobacco
Onion	Cucumber	Oats	Safflower
Lemon	Gourds	Alfalfa	Taramira
Orange	Tomato	Blue panic grass	Karnal grass
Grape	Sweet Potato	Para grass	Date palm
Peach	Sorghum	Rhodes grass	Ber
Plum	Minor millets	Sudan grass	Mesquite
Pear	Maize	Guava	Casuarina
Apple	Clover, *berseem*	Pomegranate	Tamarix
		Acacia	Salvadora

SELECTION OF CROPS

The selection of crops and cropping sequence for saline soils is of paramount importance, because crops differ in their tolerance to salinity. The adverse effects of salt stress on plant growth in saline soils could be minimized to some extent by growing salt tolerant crop or their varieties. The selection of the first crop, therefore, will depend upon the degree of soil salinity after the basic reclamation measures such as land drainage, field levelling and bunding and leaching have been implemented. Since complete reclamation may not be attained in the initial years, crops that are tolerant and can cope up with the salinity levels are preferred in the early phase of reclamation.

Improved Varieties

Crop varieties are known to differ in their tolerance to soil salinity. Latest improved varieties, released by the Central and State Varietal Release Committees for a particular agro-climatic and soil conditions, should be adopted for commercial cultivation. Improved varieties of different crops released for cultivation in North-Western zone are given in Table 6.2.

Table 6.2 Improved varieties of different crops for saline soils of the North-Western plains

Crop	Variety
Wheat	
Irrigated timely sown	HD2009, HD 2329, HD 2428, PBW 154, Raj 1972, KRL I-4, and WH 157.
Irrigated late sown	HD 2285, HD 2270, PBW 226, WH 2915 and Raj 2184.
Barley	Ratna, Jyoti, Amber, DL 88, BL 2, DL 200, K 125 and RD 2423.
Mustard	Pusa Bold, Kranti, Varuna, Pusa *Barani*,Krishna, Prakash, RH 30, PR 45, Vaibhav, Vardan and Rohini.
Pearlmillet	PHB 10, MBH 110, HHB 60, MIH 179, CM 46,PSB 8, ICMS 7704 and WC C-75.
Cotton	Laxmi, Sharda, H 777, HS 6, Pusa 31, F 414, F286, LH 886, LH 900, Bikaneri Nerma, G1, G27, AK235, AKH 4, Sangam, Amravati and MCU 7

Source: Minhas and Gupta (1992)

CROPPING SEQUENCE

In the saline soils of arid and semi-arid areas of north western India, cotton, sorghum, pearl millet, clusterbean and mothbean could be grown during summer. During winter wheat, barley, mustard, safflower and sugarcane could be grown depending upon the availability of water. The crop production as a function of soil salinity under field conditions has been studied by Sharma *et al.* (1992). It has been reported that the threshold values (Salinity upto which there is no yield reduction) for wheat, barley and mustard are 4,7 and 8 dS/m, respectively. Singh and Sharma (1993) reported that pearl millet and cotton can be grown in saline soil upto an EC_e value of 6.5 and 7.5 dS/m, respectively without significant reduction in yield.

The cultivation of crops having low evapotranspiration and /or high tolerance is one way of compensating for water deficiency. The recommended cropping sequences for saline soils are : pearl millet-barley, pearl millet-wheat, pearl millet-mustard, sorghum-wheat or barley, sorghum-mustard, clusterbean-wheat or barley and cotton-wheat or barley. Pearlmillet-wheat, pearl millet-barley, pearl millet-mustard, sorghum (fodder)-wheat and sorghum (fodder)-mustard cropping sequences are more remunerative in saline soils (Singh et al. 1992). Cotton based cropping sequences are not beneficial since the yield of the winter crops that follow cotton crop are low. In areas with water scarcity, mustard could replace wheat in the cropping sequence since its water requirement is low as compared to wheat.

MANAGEMENT PRACTICES

Establishing a good crop stand in saline soils is a challenging task. Unlike normal soils, the cultural practices for crop production in saline soils are different. An ideal package of cultural practices besides soil fertility and irrigation water management can ensure a good crop stand vis-a-vis good yield.

Cultural Practices

Due to inhospitable soil environment, germination is adversely affected. Mortality of young seedlings and poor tillering of the crops are common in saline soils. Higher seed rate with closer spacing is advisable to counter these effects (Table 6.3). For cotton, pearl millet, sorghum, wheat, barley, mustard etc. about 25% higher seed rate over the recommended seed rate for normal soils ensures good crop stand. In case of transplanted crops, the number of seedlings per hill should be increased.

Table 6.3 Recommended seed rate for different crops in saline soils

Crop	Sowing time		Seed rate (kg/ha)			
			Soil			
			Normal		Saline	
			Sowing time			
	Normal	Late	Normal	Late	Normal	Late
Wheat	I to III week of Nov.	IV week of Nov. to III week of Dec.	100	120	125	145
Barley	Middle of Oct. to mid of Nov.	Whole of Dec. to I week of Jan.	80	100	100	125
Mustard	I to III week of Oct.	IV week of Oct. to mid of Nov.	4	5	5	6
Cotton	I week of April to mid May	Mid to end of May	12	15	15	18
			(*Desi* cotton)			
			20	25	25	30
			(American cotton)			
Pearl millet	Last week of June to mid July	III week to end of July	4	5	5	6

Source : Singh et al. (1992)

Sowing/planting practices can often be modified to obtain a more favourable salt distribution in relation to seed location or growing roots (Fig. 6.3). Since salt tends to accumulate on the ridges under furrow method of irrigation, furrow planting may help in obtaining better crop stand and yield under saline conditions. Dargan and Chhillar (1973) observed that sowing of sugarbeet on one side of the ridge half way between top and bottom, opposite to the direction of the sun gave significantly higher yield than flat and ridge sowing. The salt concentration in the root zone under this method of sowing is comparatively less owing to more accumulation of salts in southern face of the ridge due to direct and intensive solar radiation on this face. Similarly, in case of sugarcane, trench method of planting gave significantly higher yield than flat sowing which could be attributed to less concentration of salts, which move to the top of adjoining ridges (Dargan et al., 1973).

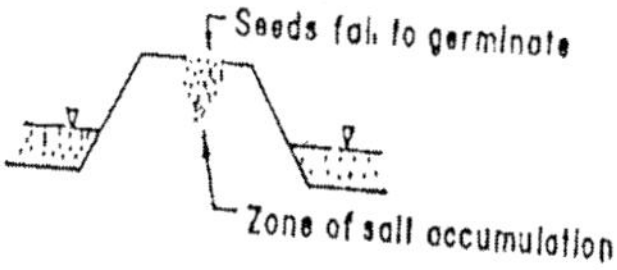

(a) Single bed, each furrow irrigated, plant fails to grow

(b) Single bed, alternate furrow irrigated, planting at the center

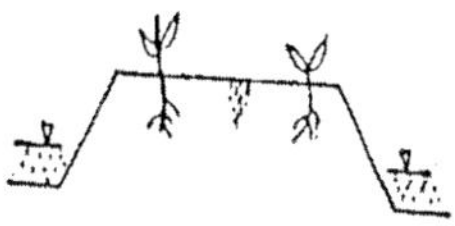

(c) Double bed, each furrow irrigated, salt accumulation at the center

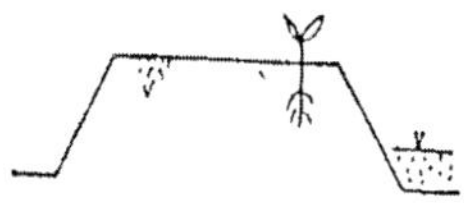

(d) Double bed, alternate furrow irrigated, planting site near the irrigated furrow

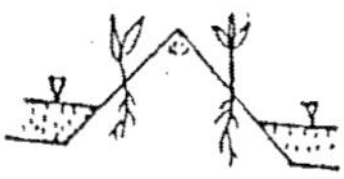

(e) Ridge and furrow, each furrow irrigated, plant on the sides of the ridge.

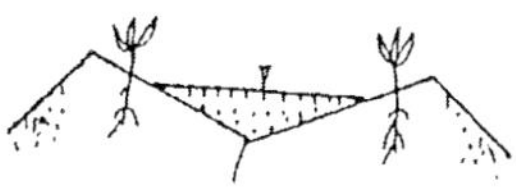

(f) Ridge and furrow, central furrow irrigated, planting near to the water surface

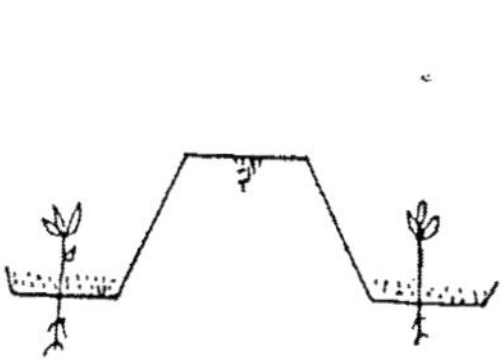

(g) Ridge and furrow, each furrow irrigated, plant in the furrow

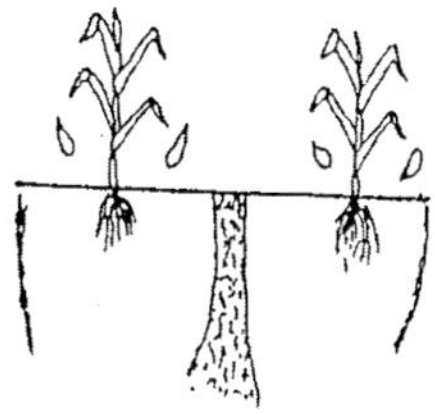

(h) Flat surface, drip irrigation, salt accumulation at fringes of the wetting front

Figure 6.3 Management of salinity through planting practices

The seeds should be sown in lines at an optimum depth with bullock or tractor driven seed drills. Seed-cum-fertilizer drill should be preferred to ensure uniform placement of seeds and fertilizers at the optimum depth. Sowing with seed drills will give a uniform stand and lead to early emergence of seedlings. If seed drill is not available, sowing should be done by *kera* or *pora* method. The method, spacing and depth of sowing of some important crops in saline soils are given in Table 6.4.

Chemical treatment of seeds improves salt tolerance of crops. In a field experiment on saline soil (pH 7.6, EC_e 10.50 dS/m) highest yield of wheat was obtained when seeds were treated with 3 per cent Na_2SO_4 solution. The yields were increased by 351.0 and 217.3 per cent in S-227 and K-65 respectively, when compared with control (Puntamkar et al., 1971). In another study, it was observed that grain yield of wheat and barley was not significantly affected when their seeds were pre-soaked in salt solution. However, pre-soaking in low salinity tubewell water resulted in more grain yield than seeds pre-soaked in NaCl or Na_2SO_4 solutions.

Pre-soaking of seeds in 50, 100, 200 and 300 mg/L of GA, IAA, IBA and NAA on the growth and development of wheat (S-227) irrigated with saline water (EC 12 dS/m, SAR 30, boron 2 and 4 mg/L) indicated that phytohormone caused differential effect in plant height, root length, shoot production and grain yield. In general, the hormones up to 200 mg/L produced significant changes in different growth attributes but affected adversely at 300 mg/L. The effect was more at higher levels. It seems that use of pre-soaked seeds in salt or hormone solutions could be a beneficial cultural management practice. However, more work is needed to be done on this aspect.

Soil Fertility Management

Adverse effect of high salinity can be alleviated to some extent by application of balanced fertilizers. Many workers have observed positive interaction between salinity and added plant nutrient thereby resulting in high yield. Crop yield generally decreases with increasing salinity but for a given salinity level the yield can be increased by judicious use of fertilizers. Reponse to added fertilizers is limited due to high salinity.

Table 6.4 Method, spacing and depth of sowing of different crops

Crop	Method of sowing	Spacing	Depth of sowing(cm)
Wheat	Line sowing with seed-cum-fertilizer drill	20-22.5cm between rows and 18-20 cm for late sown crop	5-6
Barley	Line sowing with seed-cum-fertilizer drill	20-22.5 cm between rows and 18-20 cm for late sown crop	3-5
Mustard	Line sowing (Ridge and furrow, the seed should be sown in furrow or on the north side of the ridges)	30-40 cm between lines. Thinning of plants 15-20 days after sowing is necessary to maintain 12-15 cm plant to plant spacing	3-4
Cotton	Ridge and furrow method of sowing is better in saline soils. Ridges are made about 20 cm high at the desired spacing. Cotton should be sown on the side of ridges in east-west direction, at a height of 10 cm from the base of the ridge.	60-75 cm between rows and 30-35 cm between plants in rows for American cotton. For Desi cotton, 45-65 cm between rows and 25-30 cm between plants. Extra plants should be removed two to three weeks after sowing to maintain the desired spacing between the plants.	5-6
Pearl millet	Ridge and furrow method. Sowing in the furrow in east-west direction keeps the seed zone wet and salts move to the top of the ridge (higher concentr-ation on the southern side)	60 cm between lines and 15 cm between plants. On any rainy day when the crop is 3 weeks old, helathy plants should be uprooted from places of over crowding and transplanted in the gaps.	2-3

Saline soils are generally deficient in available nitrogen and crops, therefore, invariably respond to applied N. Availability of N to crops remains low in these soils mainly due to following reasons.

i) High leaching loss of NO_3- N.

ii) Poor mineralization of organic N to inorganic forms due to high salinity and poor bacterial activity.

iii) Poor symbiotic nitrogen fixation of atmospheric nitrogen due to adverse effect of salinity on rhizobia and legumes, the host plants.

The nitrogen requirement of crops is higher in saline soils than in normal soils. High concentration of salts (KCl or K_2SO_4) inhibits nitrification and resultant NH_4-N accumulation. Hence, the plants which can absorb N only as NO_3-N will show nitrogen deficiency even if it is present in soil. Singh *et al.* (1992) have reported a significant increase in wheat yield upto 160 kg N/ha in saline soils (Table 6.5). Further they have recorded significant responses of barley, mustard, pearl millet and cotton upto 120, 100, 120 and 80 kg N/ha, respectively in saline soils. These levels are nearly 25 % higher compared to normal dose of nitrogen fertilizer for these crops in non-saline soils. Under low to medium salinity, urea gives better results than CAN and ammonium sulphate. But at high salinity levels, it is preferable to use NH_3^- containing fertilizers than fertilizers containing NH_4^+.

Table 6.5 Effect of soil salinity and nitrogen levels on grain and straw yield of wheat (Mg/ha) in saline soil

Treatment	1984-85		1985-86	
	Grain	Straw	Grain	Straw
*Soil salinity (EC_e dS/m)				
3.95	4.94	8.56	4.27	5.79
5.63	4.08	7.05	3.94	5.28
7.21	2.50	4.70	3.59	4.65
CD 5%	0.39	0.45	0.34	0.52
Nitrogen (kg/ha)				
80	3.45	5.55	3.56	4.74
120	3.84	6.86	3.88	5.21
160	4.24	7.89	4.35	5.77
CD 5%	0.39	0.45	0.34	0.52

*Mean of 1984-85 and 1985-86 at 0-20cm soil depth at sowing.

Nitrogenous fertilizer should be applied in split doses to reduce nitrogen losses through volatilization and denitrification. The required quantity of phosphorus and potash alongwith first dose of nitrogen (one third for pearl millet, wheat and barley and half for cotton and mustard) should be applied at or before sowing. The remaining quantity of nitrogen should be applied in two equal splits at first and second irrigation i.e. 25 and 45 days after sowing in pearl millet, wheat and barley and in one dose in cotton (60 days after sowing) and mustard (30 days after sowing). Under limited water supply conditions, nitrogen should be applied in two equal doses at sowing and first irrigation in wheat and barley.

Foliar application of N alongwith soil application is much economical and beneficial in saline soils. This also saves one irrigation which otherwise is to be given to dissolve and distribute effectively the top dressed fertilizer.

It has been observed by several workers that salinity adversely affects the conversion of NO_3- N into proteins thereby resulting in accumulation of inorganic N in plant parts. Thus, instead of total N content, protein-N should be the true index of N availability in saline soils. In *kharchia* variety of wheat, protein content is not affected upto EC_e of 16 dS/m. In certain parts of Haryana, Rajasthan, U.P. and M.P., the underground saline waters contain appreciable amounts of NO_3^- and sometimes become toxic to the plants. Continuous use of such waters delays maturity and adversely affects grain quality. In such areas, canal water should be used in conjunction with ground water to minimize the adverse effects.

The beneficial effects of phosphatic fertilizers on crop yield in saline soils has been reported even when soil contained high amounts of available P. This is because of the fact that soil adsorbs more P at higher electrolyte concentrations (EC) than at lower EC. Thus in a saline soil less P will be available in solution for plant growth. As salinity severely affects the root growth, the effective root absorbing surface areas diminish resulting in lower uptake of P which is a relatively immobile nutrient in soil. Moreover, uptake of P by plant roots may be adversely affected to competitive inhibition caused by excess Cl^- ions as both these anions are absorbed by the same mechanism. Thus, in saline soils phosphate fertilization would enhance the yield by directly providing the phosphorus and by decreasing the uptake of toxic elements like Cl^- and F^-. The yields of wheat and mustard increased significantly with the application of 30 kg P/ha as compared to no application of phosphorus (Singh et al., 1990; 1992) (Table 6.6). Application of phosphorus alongwith FYM has been found beneficial as this combination markedly alleviated the salt injury symptoms in crops.

Table 6.6 Effect of different levels of soil salinity and phosphorus on grain and straw yield of wheat (t/ha)

Treatments	1986-87		1987-88	
	Grain	Straw	Grain	Straw
*Soil salinity (EC_e dS/m)				
3.18	3.99	5.38	5.18	6.88
4.29	3.67	4.92	4.87	6.47
6.52	3.39	6.47	4.58	6.09
CD 5%	0.33	0.49	0.36	0.51
Phosphorus (P_2O_5 kg/ha)				
0	3.32	4.44	4.51	5.99
30	3.78	5.02	4.93	6.55
60	3.94	5.30	5.19	6.90
CD 5%	0.33	0.49	0.36	0.51

* Mean of 1986-87 and 1987-88 at 0-20 cm soil depth at sowing.

Saline soils are generally medium to high in available K but in certain situations such as where excessive leaching has taken place, K status could be low. Plant growth under high salinity levels generally proceeds with disturbed Na/K or Ca/K ratio and sometimes may show K deficiency caused due to antagonistic effect of Na, Ca or Mg. On moderately saline soils, application of potassic fertilizers may increase the crop yield either by directly supplying K or by improving ionic balance in the uptake of Na, Ca and Mg. Under high salinity conditions, it may be difficult to effectively exclude Na from the plant by use of potassic fertilizers and, therefore, effects would generally be on account of fertility status of soils on account of increased availability of the K. Application of potassic fertilizers (KCl) has been found to increase crop yields and N use efficiencies in saline soils. Potassium (K^+) and chloride (Cl^-) ions are particularly effective in osmotic adjustment by the leaves as both these ions are taken up very rapidly and can be accumulated in high concentrations in plants. For this reason, chloride salinity may be frequently less detrimental than sulphate salinity.

In general micro-nutrient (Fe, Mn, Cu, Zn) deficiencies have not been widely reported in saline soils. Fertilizer requirement for various crops to be grown in saline soils is reported in Table 6.7.

Table 6.7 Fertilizer requirement (kg/ha) of different crops in saline soils

Crop	N	P_2O_5	K_2O^*
Wheat			
Assured fertilizer supply	120-150	30-60	40-50
Under fertilizer constraints	80-100	30-40	20-25
Late sown irrigated wheat	80-100	30-40	20-25
Barley	100-120	30-40	30-45
Mustard	100-120	30-40	30-40
Cotton	60- 80	30-40	30-40
Pearl millet	100-120	30-40	30-40

* Potash should be applied on the basis of soil test values

The beneficial effects of organic/green manure as a source of nutrients and its effect on improvement of soil structure and permeability are well known. Incorporation of organic manures may have additional advantages in saline soils because:

(i) As stated previously, the N losses through volatilization of NH_3 are relatively more in salt affected soils. Thus, FYM/ green manure can serve as temporary binding agent for the ammoniacal pool of nitrogen and reduce its losses. Experiments show an increased response to N-fertilizers in the presence of organic material which suggest their role in reducing the volatilization losses and enhance N-use efficiency.

(ii) FYM has a beneficial acidifying effect on alkali soils both through the action of organic acids formed during its break down as well as due to replacement of sodium on the exchange complex with Ca+Mg contained in FYM. Therefore, addition of organic material would help reclamation process by reducing pH and exchangeable sodium in soils.

(iii) Because of small and less active microflora in saline soils, mineralization of organic nutrient fractions is comparatively less. So the retention of nutrients in organic forms for longer periods will guard against their leaching and other losses.

It may, however, be noted that application of organic/green manures as a primary source of nutrients may fail to meet all the nutritional requirements of crops. Hence, only a well mix of both inorganic and organic manures is recommended for maintaining higher yields and sustained production under saline conditions.

Irrigation Water Management

In saline soil, irrigation practices must take care of the osmotic stress which is created by the high amounts of salt present in the soil profile. Selection of an irrigation method should also consider that deep percolation losses are minimum so that there is no abrupt rise in the ground water table in these areas. Proper aeration to plant roots may be yet another factor which may modify the irrigation practices in saline environment.

In saline soils, salts often accumulate in the top few centimeters of the soil during non-crop periods. Under high water table conditions, period between the two crops or skipping of a crop may often cause excessive salt accumulation in the root zone. Under these conditions, both germination and yield of the following crop are adversely affected. A heavy pre-sowing irrigation to leach down the accumulated salts is, therefore, essential to improve germination and growth. Wherever available, the pre-sowing irrigation should be given with good quality water. Deep tillage and inversion have been found helpful in reducing salinity hazard.

Post sowing irrigation with fresh/saline water can also ensure reasonably good germination and seedling establishment as a result of reduced salt content around the seed and better moisture regime. It may, however, be important to apply water in adequate quantities as otherwise salts in the surface layer may accumulate right in the root zone. For example data from a study conducted at Hisar revealed that a heavy post-sowing irrigation (10 cm) yielded significantly more than a light irrigation (5 cm). While applying a post-sowing irrigation, it may also be ensured that there is no crusting problem as otherwise benefits of post-sowing irrigation may be nullified as a result of crust formation.

In saline soils, evapotranspiration requirement of the crops is affected by the presence of soluble salts. As the amount of salts increases, the water available to the plants creases even though the soil may appear quite moist. It is because the osmotic pressure of the soil solution increases with increase in the salt concentration and the plants are unable to extract water as readily as they can from a relatively normal soil. Light and frequent irrigations help in keeping the soil-water potential at a low level. Successful operation of field irrigation system must ensure the supply of water to the plants at the right time in appropriate quantity and quality and at the place where the plants need it for optimum growth. However, depending on the availability of water and weather conditions, the irrigation requirements of different crops may change (Table 6.8). On-farm water management should include proper land levelling and shaping, efficient design and layout of irrigation methods, scientific scheduling of irrigation under both adequate and deficient water supply conditions, irrigation management under high water table and crop planning for optimum water use.

Table 6.8 Irrigation scheduling of different crops based on the availability of water

Crops	Number and growth stages for irrigation			
	1	2	3	4
		Border irrigation		
Wheat	Between CRI and tillering (30-35)*	Booting (70-75)	CRI ** Booting Milking (110-105)	CRI, Tillering (40-45) Booting Milking
Barley	–do– (30-40)	Tillering Flowering (60-65)	Tillering, Late tillering (40-45) Milking (80-85)	CRI (20-25) Late tillering Flowering Milking

Table 6.8 contd..

	Furrow irrigation	
Mustard	Flowering (35-40)	Pre flowering (25-30) and Pod formation (55-60)
Cotton	Cotton requires four to five irrigations. The first irrigation should be given three or four weeks after sowing and the subsequent irrigations at two to three weeks interval, depending upon the nature of the soil and weather conditions. Flowering (60-65) and boll formation (80-85) are critical stages from the point of view of irrigation. Inadequate moisture during these stages leads to heavy shedding of flower buds resulting in low yield.	
Pearl millet	Generally one or two irrigations are enough. Pearl millet does not tolerate waterlogged conditions, therefore, water stagnation should not be allowed in the field for more than one day.	

* The figures in parentheses indicate number of days after sowing.
** CRI - Crown root initiation

Selection of a suitable irrigation method is of vital significance to efficient operation of an irrigation project. The irrigation methods are of three types, surface, subsurface and sprinklers. Surface method includes free flooding, check basin, border strip, furrows and corrugations. The subsurface method is used only under specialized conditions and, therefore, the choice is limited to the surface and sprinkler methods of irrigation. Drip method of irrigation is relatively a recent development. It could be beneficially applied to saline conditions. It is particularly suitable for irrigation with saline water as it helps in maintaining a zone of low salt concentration and higher moisture content resulting in favourable conditions for crop growth.

The time, frequency and quantity of irrigation water depend largely on water requirement of crops, climatic conditions, soil characteristics including texture, infiltration rate, moisture- tension relationship, irrigation methods, water quality and water table. Sorghum, pearl millet, cotton, *guar*, *moth*, wheat, barley and mustard can not tolerate standing water or a relatively wet moisture regime whereas linseed and sugarbeet will perform better in the wet moisture regime. Enough moisture should be present at the time of flowering and seed development stage. Substantial saving in irrigation water could be effected by proper scheduling of irrigation if water table remains high within certain limits and ground water is of good quality. The depth to critical water table beyond which the yields are reduced significantly, depends upon the crop, water quality and soil type. Pandey and Sinha (1972) observed that many crops can be grown without irrigation if watertable remains within a specified depth provided ground water is of good quality.

Salt and Water Profiles

One of the major reasons of variation in crop yield by different irrigation methods and for different qualities of irrigation water can be traced to the variations in salt and water status of the soil. Its distribution in space and time, during the growth period of crop further influences the yields. It is because of this reason that high frequency irrigations result in low water stress to the plants. Because moisture and salt profile and mean root zone salinity status are greatly modified by irrigation methods, salt profiles can be used to explain the differences between the irrigation methods.

Salt profiles in a furrow and ridge, and flat sowing of turnip under surface and sprinkler methods of irrigation are shown in Fig. 6.4 for a saline-alkali silty-loam soil. The higher concentration of salts up to 30 cm in the surface irrigation is evident (Yadav and Girdhar, 1977). Though the trend is almost reversed in lower layers, a 25 per cent increase in yield of turnip is obtained in case of sprinkler irrigation, mainly because of lower salinity in the 0-30 cm layer.

The seasonal variations in a saline water irrigated area are shown with mean salinity levels in the profile after the summer and winter crops (Fig. 6.5). This shows that root-zone salinity in saline water irrigated areas with sufficient monsoon rainfall is cyclic. Due to this reason soils in semi-arid regions are generally free from salt after summer which allows continuous use of saline water, if scientific planning and management is followed.

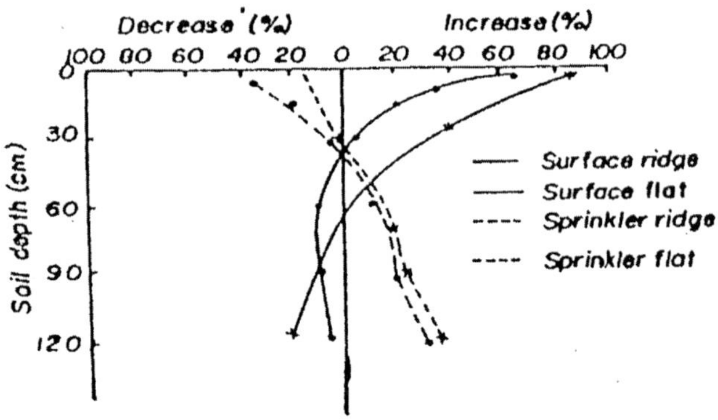

Figure 6.4 Salt profile in furrow and sprinkler method of irrigation

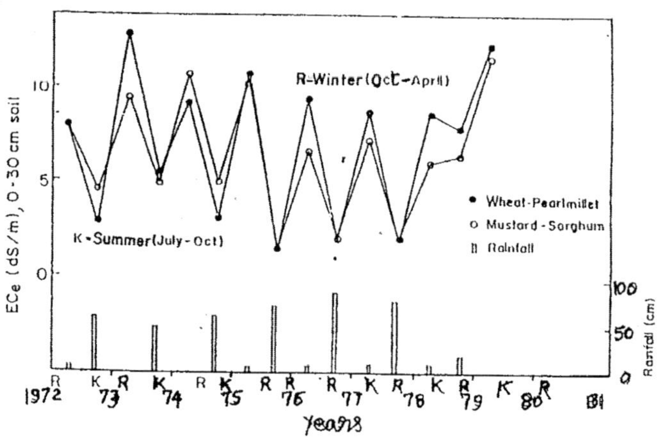

Figure 6.5 Seasonal variations in soil salinity in soils irrigated with saline water

Fallowing

As has been explained in chapter 3, saline water irrigated lands are exposed to salinisation- desalinisation cycles due to the monsoon climatic conditions. In arid conditions where rainfall is inadequate to meet the water requirement of a crop during the summer season, supplemental irrigation would be needed if a crop is grown during the season. It would require application of saline water that may render the soil more saline. Desalinisation as a result of rainfall would not be effective as well. Under these conditions, it is useful to keep the land fallow during the monsoon season so that salts accumulated during the winter season are leached out making the soil fit to grow the next winter crop. In extremely arid situations, the fallowing may have to be allowed during summer-winter-summer. It means a crop during the winter season would be grown only once in 2 years. In arid regions where fallowing during the summer season during normal and above normal rainfall years would be adequate, fallowing for summer-winter-summer may be useful in below normal rainfall years as yield of crops in saline environment would be economically unviable. At the same time soil will be salinized to a higher degree requiring more amount of rainfall in the subsequent season which may not be forthcoming.

Dhir and Bhatia (1975) observed that lands irrigated with waters below EC 5 dS/m are generally double-cropped in western Rajasthan, whereas those irrigated with water of higher salinity are used in rotation with a fallow period of varying duration but extended mostly to 18 months in between, for natural amelioration. Dhir (1977) examined salinity changes as a result of natural rainfall during the fallow phase at several locations in western Rajasthan. The study carried out over a period of 5 years, covered different soil textures. It was observed that the huge salinity build-up during the irrigation cycle is taken care of by one above-average rainfall year or two subnormal rainy seasons. Some salinity is retained at the 40-60 cm soil depth, particularly in clayloam and heavier soils, but this condition has not been found to interfere with re-use of the lands. It is also reported that it is economical to follow fallowing than to double-crop the lands in such cases.

Green Manuring

Green manuring is also one of the useful practices in the management of saline-water irrigated agriculture and therefore it should find an important place in crop rotation. Amongst several green manuring crops, the efficiency of *Ses-*

bania aculeata (*dhaincha*) has been observed to be higher than sunnhemp and *guar*. Growing of a green manuring crop in reclaimed saline soils during summer is important because of the following two reasons:

i) It provides a crop cover to the soil and therefore resalinisation of reclaimed soil is avoided. The irrigation water applied to crop also manages to keep the salt down.

ii) The crop when turned over into the soil, adds nutrients and organic matter which invariably helps in improved crop yields.

Gupta (1983) reported that *dhaincha* grows well and provides complete cover to the soil compared to sunnhemp which could not grow well in saline soils at Sampla.

Mulching

Mulching as a means to control evaporation and salinity build-up has been studied in great deal both in India and abroad. Mulches such as self mulching (soil), stubble or organic mulching and plastic mulches have been used under experimental conditions. Although, in most cases, beneficial effects have been observed, cost effectiveness has not been worked out so far. Keeping in view the cost effectiveness, mulching as a technology can not be recommended with the presently available information. However, self mulching could be an effective tool in controlling salinity build-up in the root zone with minimum cost. The self mulching usually consists of ploughing of the land surface during the intervening period between the harvest of the winter crop to the sowing of the next crop. The tillage operation thus carried out will break the capillaries, reduce the evaporation from shallow ground water table and thereby accumulation of salts in the root zone. Additionally, it will enhance rain water absorption, thus improving leaching efficiency of the rain water.

Weeds, Diseases and Pest Control

The incidence of weeds, diseases and pests in the crop grown in saline soils is not different than normal soils. As such weed control and plant protection measures including use of weedicides, fungicides and anti-biotics are practically the same as in a normal soil. Timely control and/or preventive measures to avoid weeds, diseases and pests will ensure good crop production in saline soils.

CROP PRODUCTION IN SALINE SOILS WITH SUBSURFACE DRAINAGE

Once a subsurface drainage is installed and management practices are imposed including leaching, desalinisation process becomes operative. Salt content in the root zone is reduced over the years indicating the effectiveness of subsurface drainage in land reclamation (Table 6.9). Desalinisation is quick in semi-arid regions where rainfall is on an average 500 mm or more. In these areas, a salt tolerant crop can be grown even in the first year of reclamation. Once the desalinisation is complete with provision of adequate drainage, any crop can be grown. However, in the initial years of desalinisation, crops like pearl millet, cotton, mustard, wheat and barley can be successfully grown (Rao et al., 1986). The yields of these crops increases with the decrease in soil salinity and after a few years, yields comparable to that of normal soils can be obtained (Table 6.10).

Table 6.9 Effect of subsurface tile drainage in the reduction of soil salinity (dS/m)

Soil depth (cm)	Tile spacing (m)											
	25				50				75			
	A	B	C	D	A	B	C	D	A	B	C	D
0-20	50.7	5.3	2.9	2.7	50.7	8.1	3.6	2.0	46.1	8.3	6.1	4.2
20-40	23.6	4.0	3.3	3.7	19.4	4.7	3.4	2.1	26.4	9.1	7.8	5.0
40-60	19.4	3.7	3.7	1.9	15.8	7.9	4.3	3.1	13.4	9.0	9.4	7.7
60-90	17.0	4.8	5.0	2.0	16.8	11.1	5.8	5.0	11.1	9.4	8.7	7.1
90-120	12.2	7.6	6.2	2.4	15.5	14.3	7.1	4.4	12.6	10.2	10.7	8.4
120-150	-	-	6.0	2.6	-	-	9.3	4.6	-	-	12.5	10.0
150-200	-	-	6.6	3.1	-	-	9.9	6.8	-	-	12.1	10.4

A = June 1984, B = Oct, 1984, C = Nov. 1984, D = Oct. 1988

Table 6.10 Yield of different crops (t/ha) as affected by sub-surface drainage at Sampla

Crop	1984-85			1985-86			1986-87			1987-88		
	Tile spacing (m)											
	25	50	75	25	50	75	25	50	75	25	50	75
Wheat	4.9	4.0	2.5	3.6	3.4	3.2	3.1	3.1	2.6	5.1	5.4	4.9
Barley	4.2	2.9	2.0	3.2	3.6	3.0	3.6	2.4	1.6	2.9	3.1	2.2
Mustard	2.4	2.0	0.9	0.9	1.0	1.0	1.0	0.9	0.8	–	–	–
Cotton	–	–	–	1.8	1.6	1.6	1.4	1.1	0.5	–	–	–
Pearl millet	0.8	0.7	0.4	0.3	0.3	0.2	1.4	1.3	1.1	–	–	–

ALTERNATE MANAGEMENT TECHNOLOGY

Alternate technology for the management of saline soils consists of growing salt tolerant grasses alone, afforestation or an appropriate combination of the two.

Grasses for Saline Soils

Selection of grasses for saline soils depends upon several factors such as climate, geographical conditions, the depth of water table, availability of irrigation water and quality of irrigation water. Grasses, as a rule exhibit the highest degree of salt tolerance on saline lands. Perennial forage grasses growing close to ground level are ideal from the reclamation point of view because they not only check evaporation from the soil surface but also increase the organic matter content in the soil. Bermuda grass has ben reported as tolerant grass to high salinity (Maas and Hoffman, 1977). Bernstein (1964) reported that Bermuda grass showed 10% yield reduction at EC_e 13 dSm^{-1} and 50% at EC_e 18 dSm^{-1}. On the basis of field trials conducted at the Sampla farm in Rohtak district, where there is problem of high salinity (EC_2 8dSm^{-1}), high water table (2m in the month of June) and saline ground water (EC, 22 dSm^{-1}), Kumar (1988) observed that Para grass yielded the best when good quality water was used for irrigation. Performance of Karnal grass was the best when saline water was applied to irrigate the grass. Based on the average of 4 years, it was found that Gatton panic, Karnal grass and Blue panic perform better in saline soils. Kumar and Gill (1993) while studying the effect of irrigation with highly saline water on different forage grasses in sand culture reported that grasses established better when planted by rooted slips compared to when sown with seeds. The grasses, *Panicum coloratum*, Karnal grass, Rhodes and Gatton panic were relatively less affected owing to irrigation with highly saline water. With a water of 19 dS/m, yield reduction in *Panicum coloratum* was 50%, Karnal grass and Rhodes grass about 65% and in Gatton panic about 75%. Hybrid Napier failed to grow with such waters. Anjan grass and Blue panic suffered badly even with saline waters of 12 dS/m (Kumar, 1988).

Afforestation of Saline Soils

In general, saline lands in India are more suitable for plantation of fuel than timber species. Production of fuel from saline lands depends on the ability of tree species to withstand salinity, drought and various other adverse environmental conditions. The objectives of the planting and site conditions affect the choice of species. For example *Eucalyptus camaldulensis* can tolerate saline ground water and periodic flooding, but cannot be planted close to the sea coast because of its intolerance to wind blown salt. However, *Casuarina equisetifolia* will tolerate the salt spray and can provide valuable fuel but is rarely planted inland due to its relatively low growth rate and poor coppicing ability.

In selecting fuel wood species the first requirement is tolerance of tree plant to the prevailing adverse conditions. Other desirable characteristics include potential of the plant to allow silvicultural practices, wood quality and other characteristics such as improvement of soil fertility through nitrogen fixation, its function as a wind breaker, quality in terms of food for human and animals, and its utility to extract essential oils. An ideal species will not harbour pests and diseases. Species selected to lower water table in salinity prone catchments should intercept and transpire maximum quantity of water (Burley, 1980).

Recently, rapid progress has been made in the selection of tree genotypes adapted to saline and waterlogged soils. Fuelwood has been produced on such barren lands from tree spp. of *Prosopis, Tamarix, Casuarina, Atriplex, Eucalyptus, Malaleuca, Acacia nilotica, Acacia ,tortillis, Salvadora oleoides* and *Parkinsonia aculeata* etc. Tolerance of a tree to salinity is governed by the ability of young saplings to withstand salt stress and in areas of shallow water table ability to withstand excess water stress as well. Under moderate salinity not accompanied by a shallow water table, species like *Acacia auriculiformis*, *Terminalia arjuna* and *Leucaena leucocephala* exhibited good growth but when high salinity and shallow water table co-existed, *Casuarina equisetifolia*, *Prosopis juliflora* and *Tamarix articulata* could withstand both the stresses. Tolerance of some of the tree species to soil salinity is presented in Table 6.11.

Table 6.11 Salinity tolerance of some fuelwood trees

Tolerant (EC_e 25-35 dS/m)	*Tamarix troupii, T. articulata, Prosopis juliflora, Pithecellobium dulce, Parkinsonia aculeata, Acacia farnesiana.*
Moderately tolerant (EC_e 15-25 dS/m)	*Callistemon lanceolatuy , Acacia nilotica, A. pennatula, A. tortilis, Casuarina glauca 13144, C. glauca 13987, C. obesa 27, C. glauca (FRI), C. equisetifolia (FRI), Eucalyptus camaldulensis, Leucaena leucocephala, Erescentia alata.*
Moderately sensitive (EC_e 10-15 dS/m)	*Casuarina cunninghamiana (FRI), C. Cunninghamiana (Aust), Eucalyptus tercticornis, Acacia auriculiformis, Terminalia arjuna, Pongamia pinnata, Guazuma ulmifolia, Leucaena shannoni, Samanea saman, Albizzia caribea, Senna atomaria .*
Sensitive (EC_e 7-10 dS/m)	*Syzygium cumini, S. fruticosum, Tamarindus indicia, Salix spp., Acacia deami, Albizzia guachepela, Alelia herbertsmithi, Caesalpinia eriostachyas, C. velutina, Haematoxylon brasiletto.*

Planting Technique and Irrigation Management

To get successful tree plantation in saline soils where ground waters are saline and water table remains very high, planting method should be such that the rain water is utilized to the maximum possible extent and the salt concentration in the active root zone of young plants is kept at a minimum level to minimize the adverse effect of high salinity of soil and water. Successful establishment may require different specialized site preparation and planting techniques.

Salinity and waterlogging often associated with the irrigated semi-arid regions have detrimental effect on the growth of tree plants. For establishment of tree species in waterlogged saline soils, emphasis has to be on preventing salt accumulation and improving aeration in the root-zone. Shah (1957) suggested the technique of sowing or planting *Prosopis juliflora* on high ridges instead of in pits in little Runn of Kutchh.

Field studies carried out by Tomar and Gupta (1985) on seasonal variations in salinity and soil moisture due to the fluctuating water table indicated that subsurface planting of saplings (below 30cm depth) would provide less hostile saline environment to the roots. Therefore, an experiment with ten tree species and two planting methods was initiated to evaluate the relative salt tolerance of tree species and to compare the performance of the two planting techniques for the initial establishment of the saplings (Tomar and Gupta, 1984). In the sub-surface method, saplings were transplanted in 30 cm diameter auger holes at a depth of 30 cm below the soil surface, saplings were also planted on the ridges 40 cm in height from the original soil surface. Periodical height, diameter growth and survival percentage were recorded (Table 6.12). Height, diameter and survival of tree species were invariably more in the sub-surface method than the ridge planting. Earlier results on salt build up had shown that the salt accumulation in the 30-90 cm root zone was substantially higher in the ridges than in the natural landscape soil profile. Due to higher salinity in the ridges, saplings planted on the ridges generally remain in disadvantageous position resulting in their lower survival and reduced growth. Besides high salinity, ridges also increase the surface to ground water table depths. Therefore, before the roots of the ridge planted sapling meet most of the water requirement from the less salinized transmission and phreatic zone, roots must traverse the soil depth at least equal to the ridge height. This probably explains the reason for greater mortality of saplings and their reduced growth on the ridge trench system.

Survival and growth height data of the different tree species indicate that most of the species in general performed better when planted by subsurface method (Table 6.13). The higher survival rate of saplings in furrow planting method was due to greater desalinisation of soil profile with uniform application of irrigation water.

Table 6.12 Plant height (cm) and survival percentage of tree species at different times

Tree species	Planting methods	Months after planting							
		6		27		69		107	
		H	D	H	D	H	D	H	D
Acacia nilotica	S	95	2.7	326	5.2	586	12.5	641	14.2
		(100)*		(69)		(50)		(50)	
	R	153	3.0	286	4.4	380	6.5	—	—
Prosopis juliflora	S	185	3.1	372	4.5	725	11.7	806	17.8
		(100)		(100)		(100)		(94)	
	R	196	2.9	330	3.7	594	8.4	640	13.5
		(100)		(100)		(100)		(100)	
Acacia tortilis	S	118	2.9	290	4.5	500	10.8	531	10.8
		(81)		(75)		(56)		(50)	
	R	116	1.8	157	2.3	—	—	—	—
		(69)		(38)		(0)		(0)	
Eucalyptus camaldulensis	S	110	2.2	224	3.3	593	8.9	—	—
		(81)		(50)		(38)		(0)	
	R	116	1.8	157	2.3	—	—	—	—
		(69)		(38)		(0)		(0)	
Terminalia arjua	S	99	2.2	183	4.0	414	12.7	—	—
		(100)		(81)		(50)		(0)	
	R	71	1.1	140	3.2	—	—	—	—
		(100)		(25)		(0)		(0)	
Leucaena leucocephala	S	152	2.4	277	4.7	628	9.2	691	11.7
		(100)		(69)		(63)		(38)	
	R	103	0.7	168	2.0	370	5.4	—	—
		(100)		(19)		(13)		(0)	
Eucalyptus hybrid	S	79	2.2	287	4.1	-	-	-	-
		(63)		(31)		(0)		(0)	
	R	87	1.9	181	2.6	-	-	-	-
		(69)		(31)		(0)		(0)	

* Subsurface planting method; ** Ridge-trench planting method
( ) Figures in parenthesis express percentage of survival of the tree species. H = height and D = diameter

Table 6.13 Effect of plantation methods on the survival percentage of some tree species

Tree species	Survival percentage	
	Subsurface	Furrow
Acacia nilotica	30	100
Acacia auriculiformis	40	85
Casuarina cunninghamiana	60	95
Casuarina equisetifolia	90	95
Casuarina glauca	80	90
Eucalyptus camaldulensis	50	100
Parkinsonia aculeata	75	90
Tamarindus indica	10	50
Terminalia arjuna	15	100

Tomar and Gupta (1987) compared the performance of three planting methods i.e. subsurface, ridge-trench and furrow planting. The survival and growth of 5 tree species, 27 months after transplantation revealed that furrow planted trees in general performed better than trees, planted with other methods (Table 6.14). Krepkii (1981) also observed that irrigation by furrows or in circles around the stems (basin) improved the performance of the trees, furrow irrigation being the best. On this basis, it is recommended that furrow planting of trees is most suitable for the conditions for which these studies have been made. The cost of plantation with subsurface method would be around Rs. 3600/- per ha.

Table 6.14 Effect of planting methods on survival and growth of saplings 27 months after transplanting.

Tree species	Plantation method		
	Sub-surface	Ridge	Furrow
Acacia auriculiformis	13 (1.43)*	54 (1.48)	65 (2.42)
Acacia nilotica	69 (3.21)	100 (2.85)	95 (2.89)
Casuarina equisetifolia	46 (2.13)	92 (1.11)	95 (3.00)
Eucalyptus camaldulensis	50 (2.24)	44 (1.56)	95 (3.78)
Terminalia arjuna	81 (1.83)	25 (1.36)	90 (2.00)

* Figures in parenthesis express height (m)

Cultural Practices

Irrigation is necessary for survival and growth of plants. Wherever possible, watering should be arranged particularly during the dry and hot periods. In practice, salinity effects are reduced through use of irrigation and water treatment methods that prevent excessive accumulation of salts in the root zone. Irrigation water would to some extent leach down salts out of the soil profile in case it is of good quality. Frequent irrigations with good quality water are highly beneficial for raising trees in saline soils as irrigation at short intervals provide a low salinity environment to the young roots (Singh *et al.* 1993). When good quality irrigation water is not available, saline ground water can also be used in conjunction with available fresh water. Trees of the genus *Prosopis* can be raised with saline water of EC as high as 20 dSm^{-1}.

Afforestation of salt affected community lands need provisions of adequate fresh water supplies. Lack of canal water supplies and irrigation with highly saline ground waters have often proved a serious bottleneck in the establishment of tree saplings in the initial stages of growth. This has not only hampered the implementation of afforestation but also at times in the failure of endeavors for rehabilitation of such lands.

Since waterlogged saline soils are poor in fertility and are prone to salt and moisture stresses, even the most tolerant tree species do not grow well even if all the desired management practices are followed. Therefore, it is commonly recommended to plant the trees at closer spacing than in a normal land (Singh et al., 1993). The exact spacing may depend upon the kind of the tree species. It is also observed that a dense canopy resulting from closer planting helps in checking soil salinisation at early growth stage.

The use of fertilizers to compensate for poor fertility of the soil for raising plantations on saline waterlogged soils depends on the salt concentration in the soil. The fact that the fertilizers are beneficial and allow use of saline water for irrigation has been reported in the case of normal soil for growing trees. Tomar and Yadav (1980, 1985) reported that the application of phosphatic fertilizer significantly benefited the growth of *Eucalyptus hybrid* by counteracting the harmful effect of saline water irrigation whereas nitrogen application caused harmful effects on the same forest species. More work is required to critically evaluate the response of fertilizers on the performance of tree species in saline soils.

Although the weeds are quite inhibited and do not pose a major threat to the planted saplings, weeding operation is necessary to eliminate the competition for moisture. Thus, this operation could be carried out as soon as its necessity is felt. The operation can be performed manually with the help of hand tools such as a *Khurpi.*

Agroforestry for Waterlogged Saline Lands

Such lands suffer with poor soil aeration and excess salts inhibit establishment of plants. Only those woody species which can withstand high salt content and thrive under high water table conditions should be selected for agroforestry. Tomar and Gupta (1992)reported that species such as *Acacia farnesiana, Tamarix articulata, Prosopis juliflora, Pithecellobium dulce* and *Parkinsonia aculeata* are the most tolerant to such situations. The promising grasses for planting alongwith trees are : *Aeluropus lagopoides, Brachiaria setigera, Chloris barbata, Echinocloa colonum, Dicanthium annulatum, Leptochloa fusca, Sporobolus helvolus, Cynodon dactylon, Phragmites* spp. and *Sida* spp.

Agroforestry for Saline Areas Near to Canals

Salinity problems have arisen in many canal commands because of seepage and rise in water table. Trees of the agroforestry system can act as "biological pumps" as they transpire large amounts of water. The companion halophyte shrubs/grasses could be used as ground cover and to check the upward flux of salts. Considering soil, agroclimatic and socio-economic factors, trees and companion crops can be chosen from the following :

Trees : *Eucalyptus, Casuarina, Tamarix, Populus, Prosopis, Leucaena, Acacia.*

Grasses : Kallar grass, salt grass, Alkali sacaton, cord grass, Rhodes grass, Para grass.

Bushes : Salt bush, blue bush, Kochia.

About 200 meter wide belt of *Eucalyptus* + Cord grass/ Kallar grass on both sides of the canal could be grown to intercept seepage (Singh et al., 1993).

Tree/Grass Species for Rainfed Saline Lands

About 2/3rd of global land is arid and semi-arid where crop production is dependent upon rains. Practising agriculture in such areas is always risky. However, integration of woody perennial with annual agricultural crops reduces the risk of crop failure and ensures sustainability of biomass production on such arid lands. Trees of the genus *Prosopis, Acacia* and *Parkinsonia* are promising for agroforestry in arid and semi-arid areas for food, forage and fuelwood production. Because of their low water requirement these trees do not compete with the companion crops. The yield of grasses and other herbaceous plants increased when grown in association with *Prosopis* and *Acacia* trees in water scarcity situations. The most promising agroforestry systems for arid and semi-arid areas are : *Prosopis opuntia, Prosopis cineraria - Cenchrus ciliaris, Prosopis cineraria* - millets, *Acacia albida* - millets, *Acacia tortilis* - millets. For salt affected arid and semi-arid lands *Prosopis juliflora - Opuntia ficus - indica* (Mill) is an excellent combination because both have very low water requirement and are good biomass producers.

A large area in Saurashtra region of Gujarat is saline and rainfed where only salt tolerant species of trees and grasses grow. Gururaja Rao *et al.* (1991) have identified the following important species for such lands which can be introduced into agroforestry systems for such lands.

Trees : *Salvadora spp., Prosopis juliflora, Acacia nilotica, Pithecellobium dulce, Parkinsonia aculeata.*

Shrubs : *Achyranthes aspera, Boerhavia diffusa, Digera muricata, Kochia prostrata, Suaeda fruticosa, Salsola baryosma.*

Grasses: *Aeleuropus lagopoides, Sporobolus helvolus, Cynodon dactylon, Brachiaria rammosa, Dactyloctenium aegyptium, Dichanthium annulatum, Digitaria iliaris.*

Gupta et al. (1988) innovated a novel device for the early establishment of saplings with limited water in problematic soils. The device consists of a double walled earthen pot. The diameter of the outer pot is kept aproximetely 25 cm at the top and 18 cm at the base. The diameter of the inner pot on top and at the base is kept 15 and 12 cm, respectively. Both the pots are joined together at the base allowing the basal part of the inner pot to remain almost open so that during growth the roots may easily go deep in the soil beneath (Fig. 6.6). The

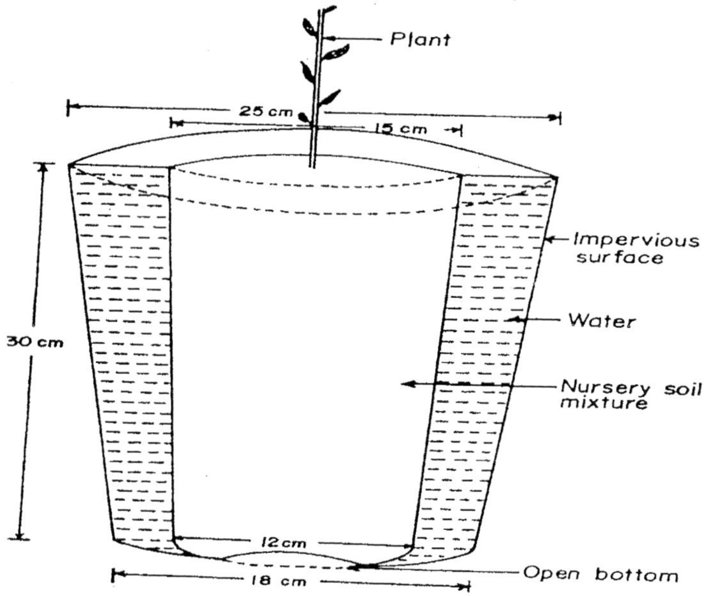

Figure 6.6 Longitudinal section of double-walled earthen pot (Jaltripti)

outer wall of the external pot is made impervious with the help of cement or some other material, to check the seepage of water outside. A tree sapling along with soil received from the nursery is placed in the inner pot. The device is fixed in a small pit so that its brim lies in line with the surface of the earth. The annular space between the two walls of the pot is filled up with water and it is covered with a polythene sheet or earthen lid with a circular hole of 15 cm diameter, so as to check the loss of water due to evaporation. Whereas in drip irrigation, the water is added to the soil in drops by gravity, in this method soil energy is used to extract the water through suction. A suction force is created by the dry soil and plant roots requiring water which draw moisture towards it from the neighbouring high moisture zones.

REFERENCES

Bernstein, L. 1964. Salt tolerance of plants. **USDA Agri. Inf. Bull. No. 283.** USDA, Riverside, California, USA.

Burley, J 1980. Selection of species of fuelwood plantation. **Comm. for Rev.** 59, 2:133-147.

Dargan K.S. and Chillar R.K. 1973. Technique for sugarbeet and vegetables in saline alkali soils. **Indian Fmg.** 22(12): 13-16.

Dargan K.S., Gaul B.L. and Abrol I.P. 1973. Growing sugarcane in saline-alkali soils. **Indian Fmg.** 23(2), 9-10,19.

Dhir R.P. and Bhatia, O.P. 1975. Use of saline waters in agriculture: I description of the system. **Ann. Arid Zone** 14 : 206-211.

Dhir, R.P. 1977. Saline water, their potentiality as a source of irrigation. **In 'Desertification and its control'** CAZRI, ICAR, New Delhi, 130-141.

Gupta, I.C., Singh, P.M., Yadav, N.D. and Sharma, B.D. 1988. Device for successful establishment of tree plants with limited water, a boon. **Invention Intelligence** 23 : 282-83.

Gupta S.K. 1983. Evaluation of leaching and resalinisation behaviour of saline soils in relation to crop growth. **Annual Rep.** CSSRI: 114-115.

Gupta, S.K. and Sharma, S.K. 1990. Response of crops to high exchangeable sodium percentage. **Irrig. Sci.** 11:173-179.

Gururaja Rao, G., Singh, R. and Bhargava, G.P. 1993. Vegetation types in Mahi Right Bank canal command, Ukai-Kakrapar command and Bhal region of Gujarat. **Tech. bull. 20. Central Soil Salinity Research Institute, RRS, Anand**. pp 52.

Krepkii, I.S. (1981). Growing trees on saline Chestnut soils of northern Kazakhstan. **Lesnoe Khozyaistvo**, 1:27-29.

Kumar, A. 1988. Performance of forage grasses in saline soils. **Indian J. Agron.** 33: 26-30.

Kumar, A. and Gill, K.S. 1993. Effect of varying salinity of irrigation waters on the yield and chemical composition of forage grasses. **Forage Res.** (Communicated).

Maas E.V. and Hoffman G.J. 1977. Crop salt tolerance: Current assessment. **J. Irrig. Drain. Div.** 103 (IR 2) : 115- 134.

Minhas, P.S. and Gupta, Raj K. 1992. **Quality of Irrgation Water: Assessment and Management.** ICAR, New Delhi. 123 p.

Misra, K.L., Porwal, A.K. and Gupta, S.K. 1990. Assessment of crop tolerance to saline irrigation waters. **Bhu Jal News.** 5(4): 35-40.

Pandey S.L., Sinha A.K. 1972. Drainage of agricultural lands. **Indian Fmg.** 22(2) : 56-58.

Puntamkar S.S., Mehta P.C., Seth S.P. 1971. Note on the inducement of salt resistance in two wheat varieties by pre-soaking with different salts of varying concentrations. **Indian J. agric. Sci.** 41 : 717-718.

Rao, K.V.G.K., Singh, O.P., Gupta, R.K., Kamra, S.K., Pandey, R.S., Kumbhare, P.S. and Abrol, I.P. 1986. Drainage investigations for salinity control in Haryana. **Central Soil Salinity Research Institute, Karnal, Bull. No. 10,** 95 p.

Singh, G.B., Singh, N.T. and Tomar, O.S. 1993. Agroforestry in Salt Affected Soils. **Tech. Bull. 17. CSSRI, Karnal.** 65 p.

Singh K.N., Sharma D.P. 1993. Influence of subsurface drainage spacings and levels of nitrogen on pearlmillet (*Pennisetum glaucum*) and cotton (*Gossypium hirsutim*) in saline soil. **Indian**

J. agric. Sci. (communicated).

Singh K.N., Sharma D.P. and Rao K.V.G.K. 1992. Technology for crop production in waterlogged saline soils. **Better farming in salt affected soils, Extn. Series,** 17, 12 p.

Singh K.N., Swarup A, Sharma D.P. and Rao K.V.G.K. 1990. Influence of subsurface drainage and phosphorus application on soil salinity and uptake of phosphorus by wheat in saline soil. **J. Indian Soc. Soil Sci.**, 38 : 511-515.

Singh K.N., Swarup A., Sharma D.P., Rao K.V.G.K. 1992. Effect of drain spacing and phosphorus levels on yield, chemical composition and uptake of nutrient by Indian Mustard (*Brassica juncea*). **Expt. Agric.** 28 : 135-142.

Sharma D.P., Singh K.N. and Rao K.V.G.K. 1992. Reuse of saline drainage water for irrigation. **Better Farming Series No. 16**, CSSRI, Karnal. 16 p.

Tomar, O.S. and Yadav, J.S.P. 1980. Effect of saline irrigation water of high salinity with and without application of fertilizers on the growth of *Eucalyptus* hybrid. **Indian J. Range Mgmt.** 2 : 121-127.

Tomar, O.S. and Yadav, J.S.P. 1985. Effect of saline water irrigation and fertilizer application on growth of tree seedlings. **Ann. Arid Zone** 24(2): 94-100.

Tomar, O.S. and Gupta, R.K. 1984. **Annual Report**, CSSRI, Karnal. 82-86.

Tomar, O.S. and Gupta, R.K. 1985. Performance of some tree species in saline soils under shallow and saline watertable conditions. **Pl.Soil** 27 : 329-335.

Tomar, O.S. and Gupta, R.K. 1987. **Annual Report,** CSSRI, Karnal. 101-103.

Tomar,O.S. and Gupta, R.K. 1992. Performance of some exotic tree species in saline soils. **Annual Report,** Central Soil Salinity Research institute, Karnal. 98-99.

Yadav, J.S.P. and Girdhar, I.K. 1977. Salt translocation and crop growth under sprinkler and surface methods of irrigation in sodic soils. **Indian J. agric. Sci.** 47:397-400.

7

RECLAMATION OF COASTAL SALINE SOILS

Coastal saline soils occur along the 6100 km long coast line of India. These occupy the deltas, lacustrine fringes, lagoons, coastal marshes and narrow coastal plain or terraces along the creeks. According to the present estimate, the coastal saline soils are spread in an area of about 2.1 mha in the coastal tracts of West Bengal, Orissa, Andhra Pradesh, Pondicherry, Tamil Nadu, Kerala, Karnataka, Maharashtra, Gujarat, Goa and Andaman & Nicobar Islands (Table 7.1 and Fig. 7.1). This estimate excludes acid sulphate soils of Kerala which are spread in an area of 0.035 mha. The problems existing in coastal saline soils may be described as:

- The lands are subjected to the influence of tidal waves and periodical inundation by tidal water.
- Shallow water table enriched with salts contributes to increase in soil salinity during winter and summer months.
- Heavy rainfall except in some parts of Gujarat.
- Poor surface and subsurface drainage conditions.
- Lack of good quality irrigation water during dry period in certain areas.

Any attempt to reclaim coastal saline soils must address to these general problems to arrive at cost effective solution.

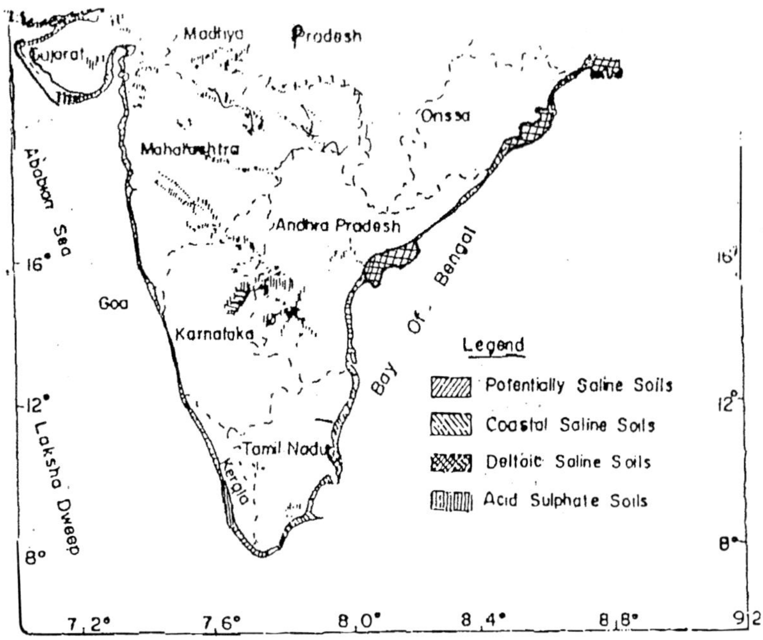

Fig. 7.1 Distribution of coastal salt affected soils in India

Table 7.1 Extent of coastal saline soils (mha)

State	Area (mha)
West Bengal	0.800
Orissa	0.400
Andhra Pradesh	0.040
Tamil Nadu	0.004
Kerala	0.016
Maharashtra	0.040
Gujarat	0.714
Karnataka and Goa	0.086
Total	2.100

Source: Yadav et al. (1979)

Physiography and Relief

Most of the coastal areas are low lying and are stretched between the eastern Ghats and the Bay of Bengal on eastern coast and between the western Ghats and the Arabian sea on western coast. Topographically, these areas have elevations less than 10 m above the mean sea level (Table 7.2). In general slope is towards the sea.

Table 7.2 Maximum and minimum elevations above mean sea level of some of the coastal places in India

State	District	Elevation (m) above sea level	
		Maximum	Minimum
West Bengal	24 Parganas	9	2
	Howrah	6	4
Andhra Pradesh	East Godavari	6	Sea level
	West Godavari	6	Sea level
	Krishna	3	Sea level
	Guntur	18	Sea level
Tamil Nadu	Madras	Sea level	Sea level
	Chinglepet	8	Sea level
	Tanjavur	5	Sea level
	Ramnathapuram	6	Sea level
Kerala	Alleppey	19	Sea level
	Ernakulam	10	Sea level
	Trichur	10	Sea level
Maharashtra	—	150	Sea level
Gujarat	Surat	10	Sea level
	Valsad	50	Sea level

Climatic Conditions

The climate of coastal region does not change much from one place to another. Except the north Gujarat coast, the area falls under hot and humid or sub-humid climatic zones. The general features of the climate are:

- Mild and short winters during December to February
- Hot and humid summers during March to May
- Rainy period during June to September with extended rainy period of October-November in the southern most peninsula.

Most of the coastal areas have equable maritime climate throughout the year. The average maximum temperature is in the range of 25 to 35° C. Only West Bengal, Orissa and Gujarat coasts have minimum temperature reaching below 20° C occasionally for about three months of December, January and February.

Almost the entire coastal area of the country excluding the north Gujarat coast receives a normal annual rainfall of more than 1000 mm. Seasonal rainfall and its percentage of the average annual rainfall in respect of different states is presented in Table 7.3. There is a wide variation of rainfall from 820 mm to 3110 mm in the entire coastal area. The coastal saline areas of Gujarat, especially Saurashtra and Kutchh receive the minimum rainfall. As a result, salinity management is a more serious problem in this region. Except South coastal Andhra Pradesh and coastal Tamil Nadu, the area receives the bulk of the rainfall from South- West monsoon (Table 7.3) from June to September. More than 80 per cent of the annual rainfall is received during this period.

Table 7.3 Seasonal distribution of rainfall (mm) at some coastal locations

Region	June-Sept.	Oct.-Nov.
West Bengal	1325 (75)*	177 (10)
Orissa	1140 (76)	180 (12)
Coastal A.P.	570 (57)	330 (33)
Tamil Nadu	340 (33)	480 (47)
Kerala	2010 (67)	550 (18)
Karnataka	2850 (87)	260 (8)
Konkan	2700 (94)	130 (5)
Gujarat Region	930 (96)	30 (3)
Saurashtra and Kutchh	450 (94)	20 (4)

*Figures in parenthesis express percentage of annual rainfall

Surface wind is generally weak. Strong or high winds are mainly in association with depressions and heavy rains which cause severe damage to crops. Cyclonic storms develop in the Bay of Bengal during October- November and strike the east coast causing heavy damage to life and property. In October, the storms largely originate between latitudes 8 and 14° N and move initially in NW direction, but a number of them finally take a curve and move in NE direction. The average number of storms or depressions during October to December based on 80 years records is four and of these only two crossed the coast (NCA, 1976). These cyclonic storms when occur together with high tides cause breaches in the protective embankments. Sometimes, even overtopping of the bunds occurs which results in inundation of large areas rendering the otherwise good coastal soils into saline soils.

GROUND WATER TABLE BEHAVIOUR

Depth to ground water table as well as its salinity, change with time during the year. Water table reaches the soil surface in August/September and falls down in May/June depending upon the initial level, rainfall distribution and evaporative demand. Variation in depth to water table and its salinity as recorded at CSSRI, Canning reveal that the water table reaches the soil surface during the monsoon season and drops to about one meter during April/May (Table 7.4). The ground water salinity is minimum during the monsoon and maximum during the summer months. The dilution due to ground water recharge during the monsoon season appears to be responsible for such a variation in the ground water salinity.

Table 7.4 Depth to water table and salinity of the ground water during different months of the year

Month	Rainfall (mm)	Depth to water table (m)	Salinity (dS/m)
January	0.0	0.97	10.8
February	11.8	0.95	13.1
March	35.4	0.98	11.2
April	25.2	1.05	10.3
May	369.6	0.53	4.6
June	287.2	0.53	6.1
July	224.6	0.17	4.4
August	224.1	0.00	2.7
September	384.4	0.00	3.9
October	191.4	0.20	4.8
November	76.6	0.56	6.1
December	173.2	0.78	9.0

SOIL CHARACTERISTICS

The characteristics of the salt affected soils found in the

coastal tracts show wide variations. The soils in general are of moderate to heavy texture and are saline in nature. Alkali soils are found only in a few pockets in Andhra Pradesh and Tamil Nadu. Heavy textured soils (silty clay to clay loam); when dry, crack heavily and look whitish due to the accumulation of salts at the soil surface. Due to heavy texture, the soils have very low water transmission characteristics and it is one of the limitations in reclaiming these soils. Except some soils of Kerala, majority of the soils have neutral soil pH all throughout the profile. In most cases, the salinity is highest in the surface horizon when sampled in April-May. Analytical data of a soil profile at village Chudamani, in the Balasore district of Orissa showed that dominant anion is Cl^- followed by $SO_4^=$. Sodium is the most dominant cation followed by Mg and Ca. Studies on the coastal soils of other states show that dominant anions are by and large Cl^- and $SO_4^=$ while in most cases $CO_3^=$ and HCO_3^- salts are absent. In acid sulphate soils of Kerala, sodium sulphate constitutes the major

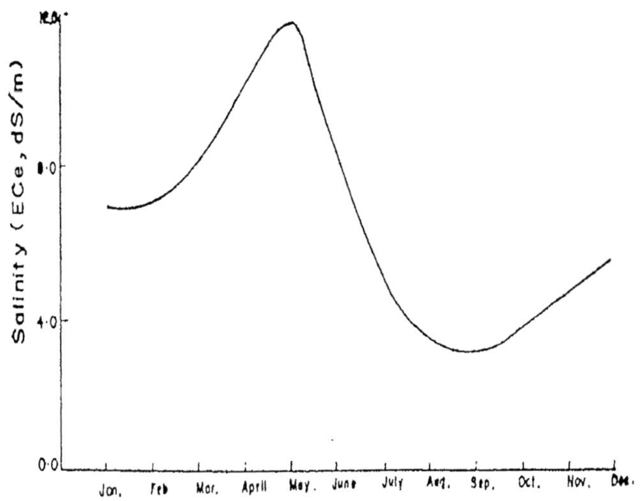

Figure 7.2 Variation of soil salinity in different months of the year

anion. The soil is acidic to neutral in reaction. A generalized account of the nature of soil salinity at different locations is reported in Table 7.5. As in the case of inland saline soils, coastal soils also follow seasonal salinisation-desalinisation cycles. A typical seasonal salinity variation in soils at Canning is presented in Fig 7.2.

Table 7.5 Nature of soil salinity at different coastal places

State	Soil texture	pH_2	ECe (dS/m)	Dominant salt
West Bengal	sc-scl	5.5-7.0	4-35	sodium chloride sodium sulphate
Orissa	cl-c	5.0-7.5	2-50	sodium chloride
Andhra Pradesh	sl-cl	6.0-8.8	0.5-17	Chlorides & sulphate of sodium
Pondichery	sl-l	6.5-8.5	1.0-50	
Tamil Nadu	sl-l	6.0-8.2	2-10	-do-
Kerala	sl-l	3.5-5.5	1-20	Sodium sulphate
Karnataka	sl-l	5.0-7.5	3-10	Sodium chloride
Maharashtra	sl-cl	7.0-8.5	4-14	-do-
Goa	sc-scl	5.0-6.0	4-15	-do-
Gujarat	sl-l	7.5-8.5	9-20	-do-

Physico-chemical Characteristics

Physico- chemical characteristics of a pedon at Upugunder (Distt Prakasam, Andhra Pradesh) and Ambalpuzha (Distt. Alleppey, Kerala) are shown in Table 7.6 and Table 7.7. The soil of the Upugunder is highly saline and depict the general characters of a typical coastal saline soil. On the other hand data in Table 7.7 is for a acid sulphate soil mostly found in Kerala. The differences in pH and relative proportions of chlorides and sulphate are noteworthy.

Development of Soil Salinity in Coastal Lands

General principles of soil salinisation have been discussed in Chapter 3. Some of the processes that are most active in coastal areas are described below:

— Sea water inundation is one of the processes by which soil salinity develops in the coastal regions. The sea water may inundate the land during high tides or during cyclones. In many cases when saline water travels to interior parts of coastal plains through rivers/creeks, farmers use this water for irrigating the crops. It enhances the process of soil salinisation.

— Use of saline ground water for crop production also causes soil salinisation in many areas.

— Wherever ground water is saline, high water table during the post monsoon season, brings up dissolved salts in the root zone through capillary rise. Salinisation process is most active during the summer seasons.

— Deposition of wind borne salts. The process is most active at Kutchh in Gujarat.

Table 7.6 Physico-chemical characteristics of a Pedon from Upgundur (Prakasam) in Andhra Pradesh

Depth (cm)	EC_e (dS/m)	pHs	Ionic composition (meq/L)							
			Ca	Mg	Na	K	CO_3	HCO_3	Cl	SO_4
0-16	53.1	7.8	91.0	267.0	547	1.2	nil	6.0	848	108
16-44	36.1	7.8	47.5	114.0	358	1.7	nil	2.5	476	30
44-100	37.2	7.7	38.0	117.5	352	1.8	nil	2.5	502	39.9

Depth	Particle size			Textural class	$CaCO_3$ (%)	OM (%)	CEC c mol (p+)/ kg	ESP
	Sand (%)	Silt (%)	Clay (%)					
0-16	43.6	30.4	24.8	l	1.2	1.0	15.3	12.4
16-44	42.6	27.6	28.4	cl	1.4	0.7	14.1	12.7
44-100	52.2	24.6	22.4	scl	0.8	0.5	10.9	11.9

COASTAL SOILS AND CROP PRODUCTIVITY

The soil salinity in the coastal belt is highly variable with the season reaching its maximum between January and May and decreasing on the onset of monsoon due to leaching, dilution and washing. In addition to salinity, a number of agro-climatic/soil conditions are unfavourable to optimum crop production. Some of the adverse features are as follows:

Table 7.7 Physico-chemical characteristics of a coastal acid sulphate soil from Ambalpuzha (Alleppey) in Kerala

Depth (cm)	EC_e (dS/m)	pHs	Ionic composition (meq/L)							
			Ca	Mg	Na	K	CO_3	HCO_3	Cl	SO_4
0-14	2.9	4.1	4.6	8.2	16.2	1.6	nil	0.5	21.2	7.3
14-32	2.0	4.2	2.5	4.5	12.0	1.5	nil	0.6	14.5	7.0
32-46	2.1	4.3	3.0	4.5	12.2	1.8		0.5	13.8	8.5
46-100	1.5	4.5	2.5	3.5	9.5	1.5	nil	0.4	10.3	6.4

Depth	Particle size			Textural class	$CaCO_3$ (%)	OM (%)	CEC c mol (p+)/ kg	ESP
	Sand (%)	Silt (%)	Clay (%)					
0-14	23.9	48.0	28.2	cl	nil	3.1	19.2	8.8
14-32	22.9	46.8	29.0	cl	nil	1.5	20.8	8.8
32-46	29.7	52.4	19.4	sil	nil	0.9	19.6	7.7
46-100	59.7	22.7	17.6	sl	mil	0.5	10.6	2.4

sl Sandy loam, cl clay loam, sil silt loam, OM organic matter

The duration of bright sunshine is an important factor affecting crop growth. With the onset of monsoon in the month of June and during most of the month of July, the average sunshine hr per day are 3 to 4. It improves to about 7 hr in the month of october but is still inadequate from crop production point of view.

The high temperature is a common feature of the coastal climate. The high temperature affects the growth of thermophyton crops such as wheat, barley, mustard, linseed etc. The relative humidity which varies from 65 to 96 % throughout the year is considered to be high. The high humidity often leads to infestation with pests and diseases.

It has been observed that when soils are submerged during the monsoon season, considerable amount of $SO_4^{=}$ salts present in these soils are transformed to sulphide salts producing H_2S gas. The initial metal sulphide content of a few mg/L increased to more than ten times after about two months. Such transformation of sulphate salts is also responsible for decrease in soil salinity during monsoon. It was observed that in saline soil of Canning, 180 to 440 cc of H_2S gas is formed/day/m^2 after two months of submergence. As H_2S at a concentration of 0.07 to 0.1 mg/L in soil solution is toxic to plants, the accumulation of H_2S in soils to toxic levels can not be ruled out. This may possibly be one of the reasons, apart from salinity, for poor rice yields in the coastal belt.

TIDES

Tides is one of the most important factor affecting the hydrology of coastal zone. The range of tides determines the shore zone and tide height would determine the height of the protective embankments to be constructed. The tides cause diurnal flooding and exposure of the shore and some parts of the coastal zone. The daily tidal range defined as the difference between the maximum and minimum height of water level is about 10m at Bhavnagar, 4m at Bombay, 0.75m at Cochin and Madras, 1.25m at Machilipatnam and 5m at Canning in West Bengal (Ahmed, 1972). The tidal range is maximum during June- December, thus affecting adversely the drainage of the monsoon water.

Height of Embankments

The coastal saline soils can be cultivated only by protecting the lands from flooding due to high tides by constructing protective embankments. The height of the dike is usually derived on the basis of ground level vis-a-vis the water level, height of the high tide expected say once in 10 years or more and the free board. Free board is usually 1 m but may be more depending upon the site condition such as wave rush, wind speed etc. Slope is usually flatter on the sea side compared to the land side. It is usually 3:1 on the sea/river side and 2:1 on the country/ land side. To circumvent the high winds and wave action a shelter belt is usually provided. In severe wave action conditions, it may be necessary to provide revetment of the slope. Where there are problems of shifting coastal sand dunes, planting of wind breaks may be necessary (Fig. 7.3).

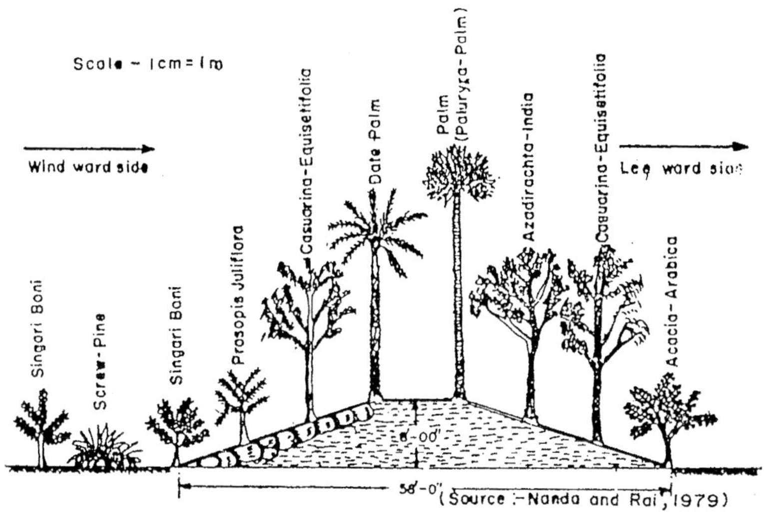

Figure 7.3 A schematic view of an embankment with shelter belt and revetment on the river side

Structural Design of Sea Dikes

Apart from the elevation of the dikes, structurally the bund must be strong enough to withstand the forces acting upon the dike. It is of no use to construct a high bund with poor structural design or vice versa. A realistic design incorporating the height and structural stability of the bund within the funds available may be a more sound proposition than to emphasize the one or the other aspect. The various modes of structural failures can be described as follows:

- Direct frontal erosion due to wave attack.
- Erosion of landward face following overtopping.
- Slipping or slumping of the landward face caused by seepage through the bank.
- Building up of water pressure at high water in pervious layers underlying the bank leading to overall instability of the bank

The structural design of the bund should also include the following:

— Slope of the bunds towards the dry and wet side.

— Seepage control.

— Selection and design of energy dissipators and revetment for the dike slope on wet side.

The design of a bund will differ from site to site and will change with the type of the soils encountered in the area.

Seepage Drain

Usually the reclamation site is at a higher elevation than the creek and therefore, water level is likely to be lower in the reclaimed area as measured from the ground surface. However, when the tide occurs, the elevation of the water level in the downward side could be higher in comparison to the ground and the ground water table in the area. Thus, it is possible that there may be a steep rise in the ground water table. The studies made in the coastal areas of Maharashtra and West Bengal (Sahu et al., 1981; Rao and Dhruva Narayana, 1979) have established that there is no sub-soil water movement with the high tide and thus it can fairly be assumed that no seepage drain along the bund in the upward direction is required in these regions. However, in case such movement is noticed, a suitable seepage drain should be designed and incorporated in the land development plan of the area.

Embankments Construction in India

Large scale embankment construction works have been reported from states along the coast. The design parameters commonly adopted in Sunderbans are based on the high tide with a 1 m of free board. Slopes of 3 : 1 on riverside and 2 : 1 on countryside are provided. Top width varies from 1 to 4 m. Wherever erosion due to wave action is expected, development of tree belt with mangrove species in the foreshore areas is undertaken. The sluice gate is a 90 cm diameter hume pipe with a flap gate opening on one side and can drain about 90 cumecs of water depending upon the head available. Multiple pipe gates are provided at locations where more discharge is expected. In Goa, however, open sluice gates have been commonly installed uptil now, but action is being taken to replace them with flap-gate type outlets.

Other Activities

Because of heavy rainfall in coastal areas, peripheral field contour bunds are helpful in demarcating the catchment, so that inflow of excess water from outside into the catchment and from one region to another are regulated. The studies conducted at Central Soil Salinity Research Institute, Regional Station, Canning indicate that for Sunderbans, the height of these peripheral bunds can be 25-30 cm (Rao and Dhruva Narayana, 1979). These bunds also help in maintaining a reasonable (10-15 cm) level of water in rice fields. Channelization of different zones of the catchment to directly drain into the outlet is also necessary.

Sluice Gates

Construction of embankments to prevent inundation of coastal soils with saline sea water would also prevent the drainage of water from inland areas unless a provision of a drain is made for the water to get out. The water through this drain will enter the area at the time of high tide and could get out from the land side to the river during low tide. The drain thus acts as a channel with two directional flow. To overcome this problem, 'outfall sluices' are constructed at the point where the drain joins a river. The outlets for this purpose are usually in the form of one way sluice gates which open only on one side. The main function of the outfall sluice is to allow free flow of drainage by gravity into the river, estuary or sea during low flood levels and prevent inflow into the drain during high flood levels. To prevent two directional flow in the drain, outfall sluices are provided with shutters. The outfall sluice is similar to a irrigation sluice valve or regulator. The vents, piers, abutments, breast wall, head wall, and aprons are designed for extreme hydraulic conditions as in the case of irrigation sluice. The size and number of vents are fixed such that the drain discharges fully into the river/estuary with a differential head of 0.30 m. The differences between various types of gates are mainly due to the shutters and their closing mechanism, so as to prevent reverse flow and allow only unidirectional flow from drain to river/estuary. Fig. 7.4 shows a typical outfall sluice with flap shutters. Proper location and realistic operational schedules would help in effective drainage of the inland areas during the monsoon season.

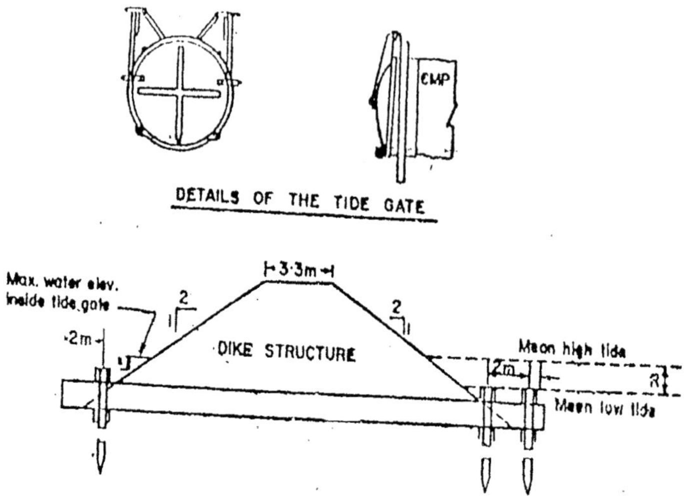

Figure 7.4 A schematic view of a sea dike and a tide or flap gate

SURFACE DRAINAGE

Excess Rain Water

Most coastal saline soils (except of the irrigated region) face a paradoxical situation. Excess rain water during the monsoon season causes high depths of water stagnation in the crop land as a result of lack of adequate drainage. During winter and summer months, there is acute shortage of good quality water for irrigation. From the water balance studies carried out for the Canning station in West Bengal, it has been observed that during June- October, as much as 450 mm of rainfall would be in excess after meeting the losses and demand during that period (Fig. 7.5). There is a big potential of storing this water for subsequent use during the winter season. The advantage of storing the excess water are : (i) to bring additional area under irrigated cultivation during dry season, and (ii) to reduce the depth of flood water on wetland rice field during wet season. It is, therefore, recommended to harvest and store excess rain water at some appropriate locations.

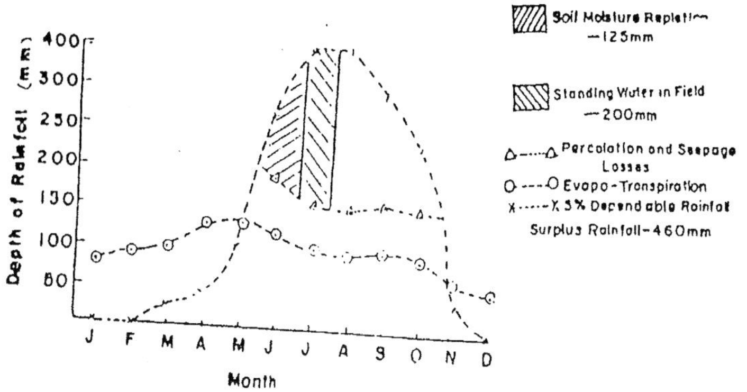

Figure 7.5 Excess water and its distribution in different months at Canning (Sunderbans, West Bengal)

Drainage Coefficient

The drainage coefficient recommended by West Bengal Flood Enquiry Commission appears to be much lower than the required value (Rao and Dhruvanarayana, 1979). The Cypress Creek's formula on the other hand is found to give quite high values for small catchments. For Sundarbans, a drainage coefficient of 37.5 mm/ day computed by simplified hydrologic procedure from the rainfall data appears to be more reasonable. Such a drainage coefficient will facilitate removal of excess rainfall, from a 5 day storm of 5 year return period within 7 days. If the losses in the field such as evapotranspiration and percolation which are about 6 mm/ day are also taken into account, then a drainage coefficient of 32 mm/ day will be quite realistic. Such site specific studies could be made to work out drainage coefficients for other sites.

Drainage of Rice Field

Most of the area during the monsoon season has standing water depths of more than 20 cm throughout the crop growth period of rice. At some places, it may be difficult to enter the fields, once the monsoon season sets in. For the improvement of the drainage system, it is suggested to construct peripheral bunds to clearly demarcate the catchment and also the various fields, so that the inflow of excess water from outside the area into the catchment and the flow from one zone to the other is regulated. This will also provide a field level control on the standing water depths. It is also recommended for channelisation of the catchment to directly route the excess rain water from different zones to the outlets. Operational schedule of the open sluice gate during the monsoon season should be so devised that gate opens more frequently and for longer periods to maintain the desired water levels in the catchment.

Critical Period for Drainage

The studies on desirable depth of submergence and the rice yields in different places of the world have revealed that the rice plant is susceptible to high water levels. The susceptibility of the plant also depends on the growth stage at which such water stagnation occurs. The period of transplantation, tillering and panicle formation stages are supposed to be more sensitive to water stagnation. Depending upon the rainfall, the crop cultivation schedules and the sensitivity of the rice seedling at the transplanting stage combine to infer that the period between first week of July and first week of August is critical for surface drainage in the coastal regions.

Storage and Recycling of Rain Water for Summer Cultivation

Possible sites for storage in coastal areas could be the following :

- Dugout farm ponds
- Main drainage channels
- Excavated natural channels
- Closed minor rivers and creeks.

Out of these alternative sites for the storage of excess water, farm ponds appear to be the most promising alternative. Usefulness of farm ponds for providing life saving irrigation to rice crop during September/ October and to grow crops like barley/leafy vegetables/chilli etc. has been established for coastal saline soils of Sunderbans in West Bengal and of Konkan region of Maharashtra. A farm pond provides complete control of the farmer on the stored water. A large number of storage structures have been constructed in the Sunderbans and elsewhere in the coastal regions. Institutional support for this activity is forthcoming because of the economic viability of the technology. Tapping the creeks could be another alternative to augment water supplies.

Design specifications for dugout ponds

Farm ponds are constructed on the lowest portion of the farm land to facilitate gravity drainage and to minimize cost of the excavation. On the basis of water balance studies at Canning (West Bengal) it has been recommended that an appropriate size to construct a farm pond would be 1/5th of the total land holding. It means a farm pond of about 0.2 ha would be required for a land holding of say 1 ha. It would provide sufficient water for irrigation to an area of 0.8 ha for crops like chilli, watermelon etc. The standard specifications for a farm pond are:

- Pond size at surface: 42.3 m x 42.3 m (1990 sq m)
- Pond depth : 3 m
- Side slope : 1:1
- Depth of water that could be utilized for irrigation: 1.5m
- Volume of earth to be excavated: 4700 m^3
- Height to which the remaining area can be raised: 50 cm.
- Fish culture: Bottom 1.5 m of the pond

A farm pond would benefit drainage in two ways. Firstly, it serves as an outlet for drainage of excess water and secondly the excavated soil is used to raise the adjoining low lying area so that run-off from adjoining lands would be minimum. Besides such structure allows growing of HYV of rice with better crop management and early sowing of winter/ summer crops (Bandyopadhyay et al., 1988).

Tapping the creeks

Creeks in coastal areas carry non- saline water during the rains and post- monsoon season but it becomes saline during the late winter/ summer. The usual practice to put up a bund

Table 7.8 Salt removal by subsurface drainage

Spacing of drains (m)	Volume of leachate collected (m)	Time (days)	Salinity (dS/m)	Salt removed (kg)	Water removed (cm/day)	Salt removed (kg/ha/day)
45	27.86	17.0	7.61	135.67	0.03	13.99
30	31.07	18.5	7.80	155.11	0.05	24.35
15	37.06	19.5	8.33	197.60	0.12	65.80

LEACHING PRACTICES

Leaching principles and practices for coastal soils are similar to that of inland saline soils. However, results of few studies carried out at some research stations in coastal areas are described and few local practices that allow leaching and crop cultivation are discussed.

Flushing

Unlike inland medium textured soils, heavy textured soils of the coastal region are amenable to flushing. In a study to establish the most appropriate method for flushing, the following treatments were tested in plots of 1000 sq.m. (40m x 25m) each. All plots were impounded with water upto 10 cm depth for 7 days.

— Impounding the land with water and draining through surface drainage wthout ploughing.

— Impounding water, ploughing and draining through surface drainage.

— Impounding water, ploughing plus puddling and draining through surface drainage.

— Impounding water, ploughing plus draining through subsurface drainage.

It was observed that,

— EC_e of soil in all treatments decreased by 4 to 5 times, while that of impounded water increased by 5 to 15 times.

— Maximum amount of salt was flushed out by ploughing plus impounding and subsurface drainage. The method was laborious, slow and time consuming.

— Surface drainage following ploughing plus impounding plus puddling was found to be quick and more effective for salt removal.

Leaching with Continuous Ponding

With continuous ponding in rice fields both during the winter and the summer months in the presence of subsurface drainage, the soil salinity decreased from 10 dS/m to about 5 dS/m to a depth of 75 cm. Thus cultivation of winter rice crop with good quality water (EC about 1.5 dS/m) would be a good practice for desalinisation provided irrigation water in sufficient quantities is available during the winter season.

Leaching with Addition of Amendments

Where the soil is fine textured and has poor transmission characteristics, the leachability of soils could be improved by addition of soil amendments. Addition of amendments like sand, rice husk, rice straw etc. are reported to be successful in improving the physical condition of soil thereby leading to efficient leaching of soluble salts.

Leaching with Rainfall

Seasonal changes in the salinity have been studied for coastal saline soils at Canning (West Bengal) where average annual rainfall is around 1800 mm. Maximum soil salinity occurred during the month of May while minimum was observed in September. It had been reported that nearly 70 per cent of the salts initially present in the soil profile could be leached during the monsoon season. Chavan et al. (1981) reported that only 20 to 50 per cent of salts could be pushed below the root zone due to rainfall (Table 7.9). In comparison to arid and semi-arid regions, leaching due to rainfall is much less in the humid regions. It could be due to high run-off as a consequence of heavy storms, poor rain water management in fields and shallow depths to ground water table.

Table 7.9 Rainfall and salt leaching in coastal saline soils (Paragaon, Maharashtra)

Annual rainfall (mm)	Salinity (% Nacl)		Reduction in salinity (%)
	May	Nov.	
2984	3.80	2.90	21.05
3248	3.60	2.50	30.55
2342	2.53	1.52	39.92
2898	3.15	1.55	41.27
2879	2.93	1.57	46.42
2026	2.57	2.04	20.62
2674	2.20	1.32	40.00
3368	2.41	1.23	48.96

Leaching and *Pokkali* cultivation

In areas where plenty of rain water is available, leaching of the salts from the soils is carried out by flushing of salts. In local parlance in Kerala, the system is known as *Pokkali* cultivation. The system consists of making mounds of about 1 square meter each of about 0.5 m in height. This operation is completed in March- April before the onset of the monsoon season. With the onset of the monsoon, during May- June, the soil is washed off the salts and the water is drained off the fields. The young seedling planted on these mounds are saved from total submergence as the monsoon rains lash the area. When after sometime these mounds are levelled, the clumps of seedlings get fixed due to weight of the soil attached to the clumps. Although *Pokkali* cultivation is crude and primitive yet it is very peculiar and unique showing the ingenuity of the people in the past. It involves scientific principles of reclaiming saline soils under these adverse conditions.

CROPPING PATTERN AND PRODUCTION PRACTICES

Cropping Pattern

The coastal areas of all the states/ union territories are mainly monocropped growing rice during the summer season. Wherever canal irrigation has been introduced, two or more crops are cultivated. Distinction between the irrigated and unirrigated areas in all the states is very clear. The cropping pattern, however changes from one state to another and within a state from one place to another. Where irrigation facilities are available, cotton,wheat, maize, chilli, groundnut, and sugarcane cultivation is being practised. Specifically in Bhavnagar district cotton and sorghum are grown during summer and wheat during winter. In arid Kutchh region, cotton or sorghum- fallow is the common crop rotation.

Agronomic and Cultural Practices

Rice is the major crop that is grown throughout the coastal belt. Compared to conventional practices, following practices help improve the yield of rice.

- Row to row and plant to plant spacing of 15cm x 15cm gave economically higher grain yield of rice.
- For high yielding semi-dwarf varieties, continuous ponding of water upto 10 cm depth gave highest yield.
- The coastal saline soils (West Bengal) are rich in potash and phosphorus. Therefore, there is no response of these fertilizers to crop yields. However, nitrogen application is essential. 80-100 kg N/ha should be applied. The nitrogen should be applied in 2-3 splits. About 75% of N should be applied upto the tillering stage and remaining at boot leaf stage.
- Slow release nitrogen fertilizers such as sulphur coated urea are useful in improving yield in deep water rice.

Barley and chilli have been identified as promising crops for the winter season. Availability of good quality water, however, is a serious constraint in getting good yields. For barley crop, 2-3 irrigations of 4 cm depth each at crown root initiation, tillering and flowering stages are sufficient for optimum yield. Chillies require about 3 L of water per plant in about 4-6 irrigations. Addition of 40 kg N/ha in full at the time of sowing resulted in highest yield from chilli plants.

Improved Irrigation Practices

Commonly, surface irrigation methods are employed for cultivation in coastal areas. Drip and/or pitcher irrigation are advantageous over the conventional irrigation (flood or plant-to- plant) methods. These methods allow more retention of soil moisture and cause more salts to move out of the crop root zone. Higher displacement of salts under drip as compared to conventional methods generally increased with increase in the amount of irrigation water. As these techniques save water, it is possible to increase winter cropping significantly in these coastal areas.

Mulching

Soil mulching with rice husk (10 t/ha) was found to improve the soil physical condition, resulting in greater intake of water into the soil under continuous submergence. Mixing rice husk with surface soil by plowing during the fallow winter periods, significantly increased the rice yield during the following monsoon season (Table 7.10) and leached salts from the soil profile.

Parur (1971) found that the application of rice straw, green manure and molasses with and without flooding had a beneficial effect on the grain yield of rice. In the loamy soils of Tamil Nadu, application of organic matter and gypsum were found to be beneficial.

COASTAL AQUIFERS AND SEA WATER INTRUSION

The aim of coastal aquifer management is to draw maximum water from the fresh water aquifer while preventing long term pollution of the aquifer. The coastal aquifers have a direct hydraulic connection with bodies of saline water or with poor saline ground water beneath the tidal waters. Because of seaward sloping nature of the coastal plains, the aquifers are recharged in inland areas. Usually fresh water recharge would depress the flow of saline ground water to the aquifer. The fresh water extends to nearly 40 times the head of the fresh water above mean sea level. However, as a result of interference by men in exploiting natural resources, the flow is reversed. Such a reversal of flow in the aquifer is termed as salt water or sea water intrusion. Following processes may be responsible for the intrusion.

— Reversal or reduction of ground water gradients as a result of over exploitation of ground water. Many areas in Gujarat have been ruined as a result of this activity.

— Destruction of natural barriers as a result of construction activities or well drilling.

Table 7.10 Grain yield of rice (t/ha) with rice husk mulching

Treatment	1973	1974	1975
Single rice crop in rainy season (T1)	3.56	4.93	3.02
T1+ additional rice crop during subsequent winter season, (T2)	3.59	5.69	4.00
Mulching with rice husk, 10 t/ha during fallow winter period, (T3e)	4.89	5.29	4.15
C.D. at 5%	0.64	0.36	0.89

Source: Bandyopadhyay and Sen, 1977

If intrusion starts, the consequences could be serious because lands once subjected to salt-water intrusion are difficult to reclaim. The situation could be corrected by adopting one or more of the following methods.

— Reduction in ground water exploitation.

— Increase in ground water recharge by artificial means.

— Construction of physical barriers.

— Construction of interceptor well for creating troughs to remove intruding inflow of sea water.

— Construction of hydraulic ridge by injected water through injection wells.

— Combination barrier usually combining the trough and the ridge barriers.

REFERENCES

Ahmed, E. 1972. **Coastal Geomorphology of India**. Orient Longman, New Delhi. 222 p.

Anonymous, 1995. Development of Doruvu technology for efficient use of fresh water floating over saline ground water in coastal sands. **Annual Progress Report : AICRP on Management of Salt Affected Soils and Use of Saline Water in Agriculture, 1994-95.** Andhra Pradesh Agricultural University, RRS, Bapatla. 15-17.

Bandyopadhyay, A.K. et al. 1988. Coastal Saline Soils of India and Their Management. **Bull No. 13. Central Soil Salinity Research Institute, Karnal.** 158 p.

Bandyopadhyay, A.K. and Sen, H.S. 1977. Introduction of a second rice crop and mulching as reclamatory management practices of saline soils of monocropped areas of south 24-Parganas, West Bengal. **J. Indian Soc. Soil Sci.** 25: 326-330.

Chavan, K.N., Chavan, A.S. and Kadrekar, S.B. 1981. Factors affecting salinity status of coastal saline soils of Maharashtra. **Proc. Sec. Conf. on Problems and Management of Coastal Saline Soils.** CSSRI Regional Res. Stat. West Bengal. 35-42 (Unpublished).

Nanda, S.S.K. and Rai, M. 1979. A note on the problems of saline lands in Orissa and their reclamation and management. **Proc. 1st Conf. on "Problems and Management of Coastal Saline Lands.** CSSRI Res. Station, Canning (W.B.)

National Commission on Agriculture (NCA), 1976. Report of the Commission, Climate and Agriculture (Part iv). **Ministry of Agriculture and Irrigation.** Govt.of India, New Delhi.

Parur, H.G. 1971. Review of research work on soil salinity and future lines of work for different zones. *In* **Soil and Water Research In India Retrospects and Prospects. Tech. Bull. No. 22**. Ministry of Agricultur, New Delhi.

Rao, K.V.G.K. and Dhrava Narayana, V.V. 1979. Hydrology and drainage conditions of coastal saline soils. **Proc. 1st Conf. on Problems and Management of Coastal Saline Soils**. CSSRI, Res. Stn. Canning (Unpublished)

Sahu, R.K., Chavan, K.N. and Kadrekar, S.B. 1981. Some hydrological observations of coastal saline soils of Konkan. **Proc. Sec. Conf. on Problems and Management of Coastal Saline Soils.** CSSRI Regional Res. Stat. West Bengal. 199-208. (Unpublished).

Yadav J.S.P. et al. 1979. Coastal Saline Soils of India. **Bulletin No. 5.** CSSRI, Karnal. 34 p.

8

SOCIO-ECONOMIC CRITERIA FOR SALINITY CONTROL

'Irrigation projects have been a part of development programmes which aim at increased and sustainable agricultural production and economic and social change in rural India. Contrary to this, it is possible to cite many examples where irrigation projects have failed to meet these objectives. Although a number of factors have been identified for the failure in a given project, a common factor has been the non-implementation of drainage plans. Inadequate drainage resulted in development of waterlogging and in gradual salinisation of the soil, with a resultant decline in land capability to produce at its optimum. Socio-economic constraints could be another reason. In order that large drainage projects are successfully implemented with minimum spread of waterlogging and soil salinisation, project plans must ensure adequate drainage and meet other technical features some of which have been discussed in previous chapters. Simultaneously, it must also meet the economic criterion and should be in tune with the social customs and beliefs of the rural masses where project is proposed to be implemented.

India now faces a stupendous task of providing artificial drainage to irrigated/saline lands which may run into several million hectares. This includes nearly 8.6 m ha area which has already gone out of cultivation. A large area which is in imminent danger of losing its production potential can also be qualified to fall in this group. Large areas may also require adequate surface drainage for improving the general productivity of lands.

The lands which have already gone out of cultivation can be reclaimed through improved drainage. Here economic aspects remain only in the background. The lands requiring investment on drainage for general improvement in the productivity, need to be governed by economic criteria.

INVESTMENTS

Investment is the allocation of current resources, with alternative productive uses, to an activity whose benefits will occur at a later date. Drainage to raise food production and for general uplift of the rural people demand large investments because drainage projects are to be implemented as a general land and water development programme. The investments are not only needed at the project level which are the primary investments but huge investments are also needed on activities which are only secondary to the project. These investments are referred to as the secondary investments. For example, if drainage has to be implemented on a large scale, investments would be needed in industries to produce drain tile and envelope materials. In most developing countries meagre funds are available for investment with conflicting demands competing with each other for larger investments. It is, therefore, necessary to strike a balance between these conflicting demands, goals to be achieved and the compelling need to raise food production.

When to Invest on Drainage

Drainage is essential for sustainable irrigated agriculture in the canal irrigated areas. A drainage component has commonly been excluded from most irrigation projects mainly to keep the investments within the available funds and to achieve high internal rate of return. It is difficult to believe that there are real financial constraints for eliminating such a vital component from the project plans. It is now known that even in technical circles, two schools of thoughts have emerged on appropriate time of implementing drainage projects within a irrigation project. According to one group, if drainage problem is expected to surface several years after the introduction of irrigation, it is wise to postpone the investments on drainage. The motive behind this approach seems to be that once a prosperous economy is established, expensive drainage schemes can be introduced. Contrary to this thinking, the other school is of the opinion that such a thinking is nothing but a economic short- sightedness. Such plans would leave the farmers in lurch after some years of euphoria. This group is in favour of implementing drainage simultaneously with the introduction of irrigation. Three major arguments of this group are:

— There is no common appropriate time limit among the various projects or lands within a project at which drainage needs to be implemented. As the appropriate time approaches, some damage to lands has already been done. In some individual cases, the damage may be severe to bear with.

— Cost of preventive drainage in the initial stages is less compared to when the damage has already been done.

— Management which exerts influence on the performance of the project can be better carried out once for all rather than in stages.

The experience of the irrigation projects in India clearly reveals that these arguments have considerable force. For example in the Chambal Command in Rajasthan, waterlogging developed seriously within first few years whereas in other projects the spread was fast to gradual. Within a project, the problem is severe at the upper end of the system and near to the canals and distributaries compared to areas which lie at the tail end. Thus, it has been difficult to judge the appropriate time in most cases. The delay in implementing drainage schemes has been responsible for large areas going out of cultivation.

ECONOMIC CRITERIA

Economic criteria come into play as more than one technically feasible designs can be prepared. On the face of it, each one would appear to be superior to the other. In order to select the most economically viable project, an economic criterion is applied to all the possible technical solutions. In most cases, one or more of the following economic criteria are evaluated.

i. Benefit-cost ratio

ii. Net present worth

iii. Rate of internal return

There can not be any hard and fast rule to which one of the above criterion should be selected for evaluating the design. For a project to be viable, benefits must exceed the cost, or the benefit cost ratio should be more than one. Therefore, economic evaluation usually calls for a comparison of the costs and benefits of all the possible alternative solutions.

Datta and Joshi (1991) assessed the financial feasibility of investments on subsurface drainage using benefit-cost ratio (B-C ratio), the net present worth (NPW) and the internal rate of return (IRR) (Table 8.1). The B-C ratio and the NPW were estimated at 1.26 and Rs. 4659 per ha, respectively. The IRR was calculated at 13.3 per cent which was sufficiently higher than the market rate of interest. The sensitivity analysis by changing the net farm income showed that the drainage for salinity control is financially feasible proposition.

Table 8.1 Feasibility of subsurface drainage

Criteria	Data sources for analysis	
	Scientific experiments	Farmers' fields
B-C ratio	3.99	1.29
NPW (Rs./ha)	7500	4659
IRR (%)	52.6	13.3

Note: Achievement is based on the yields obtained in the first 3 years

COSTS

Items of cost in a drainage project are the following:

- Land acquisition cost
- Initial investment in the project including field collector and main ditches, control works, pipes and pumps, cross drainage works, land development, roads and re-allocation of existing structure.
- Investment on replacements in future.
- Cost of operation, management and maintenance of the project.
- All other associated costs with the project such as outlays for fertilizers, pesticides and implements. This cost is particularly important when the drainage projects are planned for barren lands or when benefits are

measured in the form of increased agricultural production.

— Cost of damage to the environment of the area. For example, marine life in creeks will be severely affected with large saline effluent coming out of drainage projects.

In the cost benefit calculations, each item of cost is assigned a price. It is a accepted practice to base the calculations on a constant price level. Future inflation is neglected mainly because it is assumed that the value of money declines by that amount. If relative deviations in price or costs are expected, this fact should be reflected in the calculations.

An accounting procedure is used to convert all price on to a common base by discounting over the economic life of the project by the following relationship :

$$C = \sum_{t=1}^{n} \frac{C_t}{(1+i)} \text{ (t)} \qquad (8.1)$$

Here C is the cost at the base level, C_t is the cost in the t year of the project (t = 1,2,3...n) and i is the discount rate.

BENEFITS

The project benefits accrue due to increased production or due to damage prevented. The benefits of a drainage project may occur on account of the following :

— Increased production

— Reduction in cost of farm operations and inputs.

— Benefits obtained from timely operations.

— Advantages resulting from the introduction of new cropping patterns.

— Indirect effects such as on soil structure and hydrol ogy.

— Reduced floods, soil erosion and damage due to siltation in and outside the area.

— Benefits to public and private properties due to general lowering of ground water table.

— Indirect benefits such as employment generation and health etc.

It is very important to include any advantages that accrue as a result of improvements outside the project area. For example with a project on subsurface drainage, ground water table would be lowered in the project area resulting in an increase in the inflow of water from outside the area simultaneously decreasing the outflow. This will result in improved drainage of an area where drainage has not been implemented.

Similar to the costs, prices are also assigned to the benefits from a project. It is often a more difficult task to assign prices to benefits compared to assigning prices to costs. The discounted benefits are obtained by the following relation:

$$B = \sum_{t=1}^{n} \frac{b_t}{(1+i)} (t) \qquad (8.2)$$

where b_t is the benefits obtained in the t year and B is the total benefits at the same base level as for costs and i is the discount rate.

COMPARISON BETWEEN PROJECT AND NO PROJECT SITUATIONS

Another way to analyse the economic feasibility of a project is to compare the project benefits with a situation as if the project has not been executed. For this purpose, two approaches are usually adopted.

i. Comparison with the pre-project conditions.

ii. Comparison with the anticipated developments in the time span of the project.

The comparison with the pre-project levels is not likely to yield true benefits of the project because the spread of waterlogging and salinity in due course of time will result in more and more degradation of the land and water resources. This deterioration will not be reflected in such calculations. The true benefits can only be assessed by the second approach. The second approach is also justified because in cases where some improvements (rather than deterioration) are expected, these will be reflected in the comparison. The ultimate aim of any comparison should be to bring out the true facts and neither too little nor too much should be assigned to the project during evaluation.

To investigate the economic justification of drainage

projects that have been implemented in the Egypt, yields of the main crops were assessed by random sampling inside each project area before and after drainage. These yields have been compared with yields in undrained areas outside the project area over the same time period. The results showed that yields have increased in almost all tile drained lands. The magnitude of the increase varied with the crop, location, yield level before drainage and other factors that affect crop production. The yield data shown in Table 8.2 provide the best estimates for the effect of tile drainage on crop production in the Nile Delta Drainage Project I in different years after the project execution (Amer and Ridder, 1989).

Table 8.2 Estimated yield increase after drainage in drainage project I (Average percentage increase over the non-drained areas)

Area in feddans*	160,000	350,000	485,000	600,000
Years	72/73-76/77	72/73-77/78	72/73-78/79	72/73-79/80
Wheat	16	14	12	13
Cotton	12	13	15	15
Maize	16	17	19	16
Rice	15	–	10	14

1 feddan= 0.4 ha approx.

In the Indian context, annual cropping intensity in the drainage area was observed to increase from 73 per cent from pre-drainage level to 150 per cent in 3 years after drainage at Ismaliya in Haryana. After 3 years of the project, winter fallow declined to only 7 per cent from a pre-project level of about 60 per cent. There was substantial increase in the yield of almost all the crops, the maximum being in the case of sugarcane (Table 8.3)

Table 8.3 Average yield increase (percentage) before and after 3 years of drainage execution

Crop	Ujhana	Bhana-Bhahamna	Ismaliya
Wheat	31	4.5	89
Barley	24	12	23
Mustard	–	10	44
Sugarcane	–	–	130

Source : Datta and De Jong (1991).

DISCOUNTING AND RECLAMATION

The net benefits from a reclamation project could be more than a project designed to improve the productivity of normal land resource. However, the lag in response could result in lower discounted benefits. The priority attached to increasing food production as well as larger indirect and social benefits of a reclamation project will usually justify a drainage project for reclamation. It is now known that reclamation projects generate more employment, result in rehabilitation of the village poor and provide better environment.

SOCIAL CRITERIA

Many a times social benefits overweigh the economic benefits. Therefore, in many situations projects are undertaken even if they are economically unviable. Net benefits and the capacity to generate employment together help in deciding the viability of a project. With increasing level of employment generation, projects even with low net benefits can be approved and implemented.

World Bank also finances many schemes with low benefit-cost ratios provided such a scheme generates more employment opportunities or lead to rehabilitation of the poor. Some schemes are approved provided such schemes improve the hygienic conditions or environment of an area.

HUMAN HEALTH

Introduction of canal irrigation has often resulted in increased human health problems arising out of water borne diseases. In some cases, running water could be responsible for this situation while in others standing water could be the causative factor. Increased incidences of malaria have been reported from waterlogged areas as Anolpheles mosquitoes breed in the standing water. Documented evidence of such a situation is a report on the migration of people living in and around Amritsar in the early years of this century. The vertical drainage scheme designed for the town of Amritsar (India) in 1917 can also be quoted in this respect. The main aim of the scheme was to improve the general hygienic conditions of the area by lowering the water table in order to halt the migration of the population from the Amritsar city.

EMPLOYMENT

Another social benefit that must be included in evaluating a project is its capacity to generate employment opportunities. Drainage projects, on this account alone have good chances of approval as these projects generate large employment opportunities. The projects will not only employ trained engineers but also a large number of skilled and unskilled labour will be engaged. At the field level, it will generate work for the farm family. Capacity to generate employment opportunity at the field level due to a reclamation project for an alkali land is given in Table 8.4. In projects related to reclamation of saline soils, the employment generation capacity is at a still higher level (Table 8.5). In subsequent years, the employment generation will be on account of land cultivation and due to increased output expected from improvements in soil and water resources of the region.

Table: 8.4 Capacity for employment generation at field level due to reclamation of 1 hectare of barren alkali land

Year	Particulars	Number of		
		Man days	Bullock days	Tractor hours
I	Reclamation	30.56	10.23	13.26
	Rice	70.30	0.95	–
	Wheat	47.09	7.51	3.44
II	Rice	86.65	6.86	2.97
	Wheat	34.90	5.64	3.77

Source: Singh (1980)

Table : 8.5 Employment generation for implementing sub-sur face horizontal tile drainage system

Spacing (m)	Installation of laterals and collections	Construction of		Total
		Junction boxes	Sump	
25	289.95	4.61	13.77	303.33
50	169.75	2.56	13.77	186.08
75	112.15	1.54	13.77	127.81
100	102.90	1.54	13.77	118.21

Source: Joshi et al. (1987)

INTER SECTORAL TERMS OF TRADE

Resource mobilization from one sector to another is an indication of strengthening inter-sectoral terms of trade and economic development. Total cost involved in installing a sub-surface drainage system can be divided into three different sectors. The data reveal that share of the industrial sector dominates the total cost of drainage being as high as 59 per cent (Table 8.6). The share of the rural based non-farm resources occupy the second place with 20.77 per cent. The data suggest that the drainage projects of this nature will strengthen both the industrial and the agricultural sector.

Table: 8.6 Share of different sectors in total cost of drainage

Sector	Share in total cost (per cent)
Industrial	59.10
Rural based non-farm resources	20.77
Rural labour	20.13
Total	100.00

Source: Joshi et al. (1987)

COST RECOVERY

To implement a drainage project is a costly exercise. The cost may vary from Rs. 15,000 to Rs. 25,000 per hectare. Therefore, when a drainage project is proposed for an area, the government has to make the initial investments from its sources. However, the question arises 'who will pay the costs ultimately and what portion of the costs should be recovered? Question of subsidizing the costs may also crop up as most farmers are with small land holdings. Principally, it is justified to recover the whole cost from the beneficiary farmers. At the same time, it is also necessary that the recoveries should not exceed the net benefits, a particular group of farmers is likely to receive from the project. The net benefits of a land drainage project depend upon several factors such as given below :

i. elevation of the land in relation to drainage

ii. proximity of main drain to the land

iii. distance from the natural outlet

iv. thoroughness of drainage

v. soil fertility

vi. variations in prevailing grades

It is abundantly clear that in a drainage project, net benefits can be highly variable. It is, therefore, necessary to know this variability before the costs are apportioned among the beneficiaries. To evaluate this variability at the level of individual farmers, may be quite complex and may be beyond the reach of human resources available in the project. Therefore, it is advisable to divide the area into blocks of 15-20 ha for surveying the variability. Other method of distributing costs among the beneficiaries as listed below are also prevailing.

i. Arbitrary assessment

ii. Uniform cost/ha

iii. Assessment on the basis of increased land value

iv. Assessment by classification

v. Assessment on the basis of 100 points

In the last case, maximum benefits are given an index of 100 points. Other areas with relatively less benefits are awarded points on this basis. These points are then used to distribute the costs.

It may also be reasonable to argue that general benefits to the public property should be clearly worked out and should not be charged to the farmers. In other words, it means that it is not advisable to recover 100 per cent cost from the farmers. As per policy being adopted in the Egypt, there is no recovery from the farmer for the first 3 years after the drains have been installed. Thereafter, the actual cost is recovered in instalments over a period of 20 years. The recovery is made alongwith general revenue for the supply of irrigation water.

SUBSIDY

There is ample justification for granting subsidy to the farmers on the cost of drainage. It should either be adjusted in the recovery rates or should be granted during the project implementation in case farmers undertake field drainage works at their own level. The following factors justify the subsidy.

i. The lands have been damaged due to canal irrigation for which farmers have paid to the state exchequer. They should not be penalized for technical faults in the planning of a project.

ii. Drainage projects have many indirect social benefits which will be shared by all. As persons who do not own lands will be sharing the benefits but not the cost, the subsidy to the farming community is justifiable.

REFERENCES

Amer, M.H. and de Ridder, N.A. (Eds.), 1989. **Land Drainage in Egypt**, Drainage Research Institute, Cairo, Egypt.

Datta, K.K. and de Jong, C. 1991. The effect of subsurface drainage on the farm economy. **Report Int. Inst. Land Recl. and Mgmt. Wageningen,** The Netherlands.

Datta, K.K. and Joshi, P.K. 1991. Cost and benefits of subsurface drainage for the control of soil salinity and waterlogging. **Better Farming in Salt Affected Soils No. 20**. C.S.S.R.I., Karnal.

Joshi, P.K., Singh, O.P., Rao, K.V.G.K. and Singh, K.N. 1987. Subsurface drainage for Salinity control-an economic analysis. **Indian J. Agric. Econ**. 42: 198-206.

Singh, P. 1980. An appraisal of the economics of reclamation of alkali soils. **Proc. Int. Symp. Salt Affected Soils,** Karnal. 555-61.

ANNEXURE - I

A. Programme listing for frequency analysis

```
PROGRAMME FOR FREQUENCY ANALYSIS OF RAINFALL
TESTS NORMAL, LOGNORMAL, SMEMAX, GUMBLE DIST.
CALCULATES INVERSE OF GAUSSIAN NORMAL
DISTRIBUTION
EVALUATES CO-RELATION COEFFICIENTS
RAINFALL IN mm(INPUT), OUTPUT IN mm
NC=NUMBER OF CASES
NOB=NUMBER OF OBSERVATION IN EACH CASE
X(I)= RAINFALL THE VALUES SHOULD BE EQUAL TO NOB
-----------------------------------------------
DIMENSION  Y(100), X(100), RKS(100), AK(100), Z(100)
DIMENSION YM(100), YP(100), TP(100), XP(100), AKP(100),
1P (100)
DIMENSION WA(100), XG(100), RK(100),TN(100),TL(100),TO
(100)
DIMENSION PP(100), TO1(100), WB(100), X1(100), 1T1(100),
X2(100)
DIMENSION TL1(100), TP1(100), XP1(100)
DIMENSION WB1(100),XD(100),ZDL(100),XDL(100),XD1
(100),1XDL1(100)
REAL KURT, KURTL, KURTZ, KURTT, KURTX
DATA  (P(I),I=1,11)/.99,.95,.90,.80,.50,.20,.10,.04,.02,.01,
1.005/
DATA  (AKP(I),I=1,11)/-2.33,-1.65,-1.28,-.845,0.0,.845,1.28,
```

```
        11.75,2.055,2.33,2.575/
        CALL FILEI
        CALL FILEO
        READ(5,*)NC,NOB
        DO 10 K=1,NC
5       READ(5,*) (X(I),I=1,NOB)
        DO 18 I=1,NOB
18      X1(I)=X(I)
        WRITE(6,100)
100     FORMAT(10X,'FREQUENCY ANALYSIS OF RAINFALL')

        SORT OUT THE DATA IN DECENDING ORDER

7       LIMIT=NOB-1
        LASTAW=NOB
        DO 900 J=1,NOB
        IF(LASTAW.EQ.1) GO TO 102
        LASTAW=1
        DO 901 I=1,LIMIT
        IF(X1(I).GE.X1(I+1)) GO TO 901
        TEMP=X1(I)
        X1(I)=X1(I+1)
        X1(I+1)=TEMP
        LASTAW=I
901     CONTINUE
        LIMIT=LASTAW-1
900     CONTINUE
102     CONTINUE

        CALCULATES OBSERVED FREQUENCIES

        ANOB=FLOAT(NOB)
        DO 6 I=1,NOB
        S=FLOAT(I)
        Y(I)=ALOG(X(I))
        RK(I)=S/(ANOB+1)
6       CONTINUE
```

```
      CALCULATES INVERSE IO GAUSSIAN NORMAL
      DISTRIBUTION

      WRITE(6,101)K
101   FORMAT(10X,'EXAMPLE=',I3)
      WRITE(6,1002)
      DO 17 I=1,NOB
      RKS(I)=ABS(2.0*RK(I)-1)
      S1=2.0/(1-RKS(I))
      S2=SQRT(ALOG(S1*S1))
      ANUM=(0.010328*S2+0.802853)*S2+2.515511
      DIN=((0.001308*S2+0.189269)*S2+1.432788)*S2+1
      AK(I)=S2-ANUM/DIN
      IF(RK(I).LT.0.5) THEN
      AK(I)=AK(I)
      GO TO 17
      ELSE
      END IF
      AK(I)=-AK(I)
17    WRITE(6,1003)I,X1(I),RK(I)
      SUM=0
      SUML=0
      SUMZ=0
      SUMSX2=0
      SUMSX3=0
      SUMSX4=0
      SUMSL2=0
      SUMSL3=0
      SUMSL4=0
      SUMSZ2=0
      SUMSZ3=0
      SUMSZ4=0
      A1=1/(ANOB-1)
      A2=ANOB/((ANOB-1)*(ANOB-2))
      A3=ANOB**2/((ANOB-1)*(ANOB-2)*(ANOB-3))
      CALCULATES STAND.DEV.,SKEWNESS COEFF &
      KURTOSIS OF ORIGINAL SERIES
```

```
      DO 8 I=1,NOB
      SUM=SUM+X(I)
8     CONTINUE
      AVGX=SUM/NOB
      DO 9 I=1,NOB
      SUMX=X(I)-AVGX
      SUMSX2=SUMSX2+SUMX**2
      SUMSX3=SUMSX3+SUMX**3
      SUMSX4=SUMSX4+SUMX**4
9     CONTINUE
      SIGMA=SQRT(ABS(SUMSX2/(NOB-1)))
      SKEW=A2*SUMSX3/SIGMA**3
      SKEW1=SKEW
      KURTX=A3*SUMSX4/SIGMA**4
      ASKEW=ABS(SKEW)
      CVT=SIGMA/AVGX
301   WRITE(6,1000)
      WRITE(6,1015)AVGX,SIGMA,SKEW1,KURTX,CVT

      CALCULATES STAND.DEV., SKEWNESS & KURTOSIS FOR
      LOGNORMAL SERIES

      DO 15 I=1,NOB
      SUML=SUML+Y(I)
15    CONTINUE
      AVGLX=SUML/NOB
      DO 11 I=1,NOB
      SUMLL=Y(I)-AVGLX
      SUMSL2=SUMSL2+SUMLL**2
      SUMSL3=SUMSL3+SUMLL**3
      SUMSL4=SUMSL4+SUMLL**4
11    CONTINUE
      SIGMAL=SQRT(ABS(SUMSL2/(NOB-1)))
      SKEWL=A2*SUMSL3/SIGMAL**3
      KURTL=SUMSL4/SIGMAL**4
      ASKEWL=ABS(SKEWL)
      AKURTL=ABS(KURTL)
```

```
      WRITE (6,1034)
      DO 555 I=1,11
      TN(I)=AVGX+AKP(I)*SIGMA
      TO(I)=AVGLX+AKP(I)*SIGMAL
      TL(I)=EXP(TO(I))
      PP(I)=1/P(I)
      WRITE(6,1018)P(I),PP(I),TN(I),TL(I)
555   CONTINUE

      CALCULATES RAINFALL AT OBSERVED PROBABILITIES

      DO 750 I=1,NOB
      WB1(I)=AVGX+AK(I)*SIGMA
      TO1(I)=AVGLX+AK(I)*SIGMAL
      TL1(I)=EXP(TO1(I))
750   CONTINUE
      CONVERTS SERIES SMEMAX
      XME=(ANOB+1)/2.0
      IM=(NOB+1)/2
      ZM=FLOAT(IM)
      IF(XME.EQ.ZM) THEN
      XM=X1(IM)
      GOTO 700
      ELSE
      ENDIF
      XM=(X1(IM)+X1(IM+1))/2
700   XS=X1(NOB)
      XL=X1(1)
      A=XM-XS
      B=XL-XM
      DO 16 I=1,NOB
      IF(X1(I).LE.XM) THEN
      Z(I)=50.0*(1-(XM-X1(I))/A)
      GO TO 16
      ELSE
      END IF
```

```
      Z(I)=50.0*(1+(X1(I)-XM)/B)
16    CONTINUE

CALCULATES STAND.DEV., SKEWNESS COFF. KURTOSIS OF
SMEMAX SERIES

      SUMZ=0
      SUMZ2=0
      SUMZ3=0
      SUMZ4=0
      DO 21 I=1,NOB
21    SUMZ=SUMZ+Z(I)
      AVGZ=SUMZ/NOB
      DO 22 I=1,NOB
      SUMZZ=Z(I)-AVGZ
      SUMZ2=SUMZ2+SUMZZ**2
      SUMZ3=SUMZ3+SUMZZ**3
      SUMZ4=SUMZ4+SUMZZ**4
22    CONTINUE
      SIGMAZ=SQRT(ABS(SUMZ2/(NOB-1)))
      SKEWZ=A2*SUMZ3/SIGMAZ**3
      KURTZ=A3*SUMZ4/SIGMAZ**4
      WRITE(6,1040)
      DO 23 I=1,11
      XD(I)=AVGZ+AKP(I)*SIGMAZ
      IF (XD(I).LE.50.0) THEN
      XDL(I)=XM+A*(XD(I)/50-1)
      GO TO 51
      ELSE
      END IF
      XDL(I)=XM+B*(XD(I)/50-1)
51    WRITE(6,1033)P(I),PP(I),TN(I),XDL(I)
      23 CONTINUE
      CALCULATES PROBABILITIES WITH SMEMAX
      DO 24 I=1,NOB
      XD1(I)=AVGZ+AK(I)*SIGMAZ
      IF (XD1(I).LE.50.0) THEN
```

```
        XDL1(I)=XM+A*(XD1(I)/50-1)
        GO TO 24
        ELSE
        END IF
        XDL1(I)=XM+B*(XD1(I)/50-1)
24      CONTINUE
        GUMBEL DISTRIBUTION
        C=0.57722
        ALPHA=(SQRT(6.0)*SIGMA)/3.1422
        AVGG=C*ALPHA-AVGX
        WRITE(6,1019)
        DO 701 I=1,11
        WA(I)=ALOG(-ALOG(1-P(I)))
        XG(I)=-(WA(I)*ALPHA)-AVGG
        WRITE(6,1020)P(I),PP(I),TN(I),XG(I)
701     CONTINUE
        DO 740 I=1,NOB
        WB(I)=ALOG(-ALOG(1-RK(I)))
        WB(I)=-(WB(I)*ALPHA)+AVGG
740     CONTINUE
        CORRELATION COEFFICIENT
        INDEX=1
780     SUMX1=0
        SUMX1=0
        SUMWB=0
        SSUMX1=0
        SSUMWB=0
        SSX1WB=0
        DO 800 I=1,NOB
        SUMX1=SUMX1+X1(I)
        SUMWB=SUMWB+WB1(I)
800     CONTINUE
        AVGX1=SUMX1/NOB
        AVGWB=SUMWB/NOB
        DO 850 I=1,NOB
        SSUMX1=SSUMX1+(X1(I)-AVGX1)**2
```

```
      SSUMWB=SSUMWB+(WB1(I)-AVGWB)**2
      SSX1WB=SSX1WB+(X1(I)-AVGX)*(WB1(I)-AVGWB)
850   CONTINUE
      R=SSX1WB/SQRT(SSUMX1*SSUMWB)
      IF(INDEX.EQ.1)WRITE(6,1023)R
      IF(INDEX.EQ.2)WRITE(6,1024)R
      IF(INDEX.EQ.3)WRITE(6,1025)R
      IF(INDEX.EQ.4)WRITE(6,1026)R
      IF(INDEX.GT.3) THEN
      GO TO 10
      ELSE
      END IF
      IF (INDEX-2) 25,26,27
25    DO 860 I=1,NOB
860   WB1(I)=TL1(I)
      INDEX=INDEX+1
      GO TO 780
26    DO 870 I=1,NOB
870   WB1(I)=XDL1(I)
      INDEX=INDEX+1
      GO TO 780
27    DO 880 I=1,NOB
880   WB1(I)=WB(I)
      INDEX=INDEX+1
      GO TO 780
      10 CONTINUE
1002  FORMAT(/9X,'I',10X,'X1(I)',11X,'RK(I)'/)
1003  FORMAT(5X,I5,5X,F11.4,5X,F10.4)
1033  FORMAT(5X,F10.4,5X,F10.4,5X,E15.8,5X,E15.8)
1034  FORMAT (//9X,'PROBABILITY',2X,'RETURN PERIOD', 2X,'
      NORMAL', 114X,'LOGNORMAL'/)
1040  FORMAT(/9X,'PROBABILITY',2X,'RETURN PERIOD', 2X,'
      NORMAL', 114X,'SMEMAX'/)
1015  FORMAT(1X,3F15.4,F10.4,F15.4)
1018  FORMAT  (5X,F10.4,5X,F10.4,5X,E15.8,5X,E15.8)
1000  FORMAT  (//10X,'AVG',12X,'SIGMA',10X,'SKEW',6X,'
```

```
      1KURT',11X,'CVT')
1005  FORmat (//5X,F10.5,5X,F10.6,5X,F15.4,5X,F10.4)
1007  FORMAT(/5X,'AVGZ=',F15.4,5X,'SIGMAZ=',F15.4,5X,'
      SKEWZ='1,F15.4,2X,'KURTZ',F10.4,2X,'ALEMDA',F15.4)
1019  FORMAT(/9X,'PROBABILITY',1X,'RETURN
      PERIOD',3X,'NORMAL',10X,'1GUMBLE'/)
1020  FORMAT(5X,F10.4,5X,F10.4,5X,E15.8,5X,E15.8)
1023  FORMAT(/2X,'COR.COEFF. FOR NORMAL DIS.=',F10.4)
1024  FORMAT(/2X,'COR.COEFF. FOR LOGNORMAL DIS.=',F7.4)
1025  FORMAT(/2X,'COR.COEFF. FOR SMEMAX DIS.=',F10.4)
1026  FORMAT(/2X,'COR.COEFF. FOR GUMBEL DIS.=',F10.4///)
      STOP
      END
      SUBROUTINE FILEO
      CHARACTER FNAME*40
      WRITE(*,11)
11    FORMAT(5X,' TYPE THE NAME OF OUTPUT FILE      ')
      READ(*,10)FNAME
      OPEN(6,FILE=FNAME )
0     FORMAT(A40)
      END
      SUBROUTINE FILEI
      CHARACTER FNAME*40
      WRITE(*,11)
11    FORMAT(5X, 'TYPE THE NAME OF INPUT FILE')
      READ(*,10)FNAME
      OPEN(5, FILE=FNAME )
10    FORMAT(A40)
      END
```

b. Sample input file for frequency analysis

```
7 24
138 110 111 101 230 93 290 308 95 187 125 139 79 111 119 104
53  176 114 178 76 44 189 134
158 166 151 172 298 110 342 504 189 342 215 163 148 158 131
136 96  195 125 277 88 75 280 164
163 185 211 192 308 162 383 632 252 372 248 167 153 173 161
```

150 96 243 192 291 100 107 295 202

177 241 251 200 313 179 491 659 338 393 269 179 178 199 165
194 99 248 258 309 129 142 300 223

193 256 251 230 315 196 532 660 364 413 269 185 200 232 168
252 99 250 270 339 141 143 307 236

196 328 261 257 316 205 553 664 366 498 269 215 208 248 169
265 110 250 282 361 145 147 322 266

220 373 294 290 317 209 568 664 367 528 296 238 216 252 190
284 118 251 322 375 150 147 326 293

c. Smaple output file for frequency analysis (A part of the file is reproduced to cover only one case)

FREQUENCY ANALYSIS OF RAINFALL

EXAMPLE= 1

I	X1(I)	RK(I)
1	308.0000	0.0400
2	290.0000	0.0800
3	230.0000	0.1200
4	189.0000	0.1600
5	187.0000	0.2000
6	178.0000	0.2400
7	176.0000	0.2800
8	139.0000	0.3200
9	138.0000	0.3600
10	134.0000	0.4000
11	125.0000	0.4400
12	119.0000	0.4800
13	114.0000	0.5200
14	111.0000	0.5600
15	111.0000	0.6000
16	110.0000	0.6400
17	104.0000	0.6800
18	101.0000	0.7200
19	95.0000	0.7600
20	93.0000	0.8000
21	79.0000	0.8400
22	76.0000	0.8800

23	53.0000	0.9200
24	44.0000	0.9600

AVG	SIGMA	SKEW	KURT	CVT
137.6667	66.4835	1.2010	4.5440	0.4829

PROBABILI TY	RETURN PERIOD	NORMAL	LOGNORMAL
0.9900	1.0101	-0.17239925E+02	0.41504660E+02
0.9500	1.0526	0.27968865E+02	0.57123827E+02
0.9000	1.1111	0.52567768E+02	0.67966932E+02
0.8000	1.2500	0.81488096E+02	0.83375430E+02
0.5000	2.0000	0.13766667E+03	0.12399917E+03
0.2000	5.0000	0.19384524E+03	0.18441635E+03
0.1000	10.0000	0.22276557E+03	0.22622461E+03
0.0400	25.0000	0.25401282E+03	0.28211113E+03
0.0200	50.0000	0.27429032E+03	0.32556712E+03
0.0100	100.0000	0.29257327E+03	0.37045940E+03
0.0050	200.0000	0.30886173E+03	0.41564336E+03

PROBABILI TY	RETURN PERIOD	NORMAL	SMEMAX
0.9900	1.0101	-0.17239925E+02	0.35295286E+02
0.9500	1.0526	0.27968865E+02	0.59013378E+02
0.9000	1.1111	0.52567768E+02	0.71918809E+02
0.8000	1.2500	0.81488096E+02	0.87091410E+02
0.5000	2.0000	0.13766667E+03	0.11667071E+03
0.2000	5.0000	0.19384524E+03	0.19452065E+03
0.1000	10.0000	0.22276557E+03	0.23459726E+03
0.0400	25.0000	0.25401282E+03	0.27789837E+03
0.0200	50.0000	0.27429032E+03	0.30599811E+03
0.0100	100.0000	0.29257327E+03	0.33133387E+03
0.0050	200.0000	0.30886173E+03	0.35390573E+03

PROBABILITY	RETURN PERIOD	NORMAL	GUMBEL
0.9900	1.0101	-0.17239925E+02	0.28602004E+02
0.9500	1.0526	0.27968865E+02	0.50887143E+02
0.9000	1.1111	0.52567768E+02	0.64525735E+02
0.8000	1.2500	0.81488096E+02	0.83087426E+02
0.5000	2.0000	0.13766667E+03	0.12674635E+03
0.2000	5.0000	0.19384524E+03	0.18548844E+03
0.1000	10.0000	0.22276557E+03	0.22438083E+03
0.0400	25.0000	0.25401282E+03	0.27352145E+03
0.0200	50.0000	0.27429032E+03	0.30997678E+03
0.0100	100.0000	0.29257327E+03	0.34616291E+03
0.0050	200.0000	0.30886173E+03	0.38221701E+03

COR.COEFF. FOR NORMAL DIS. = 0.9419

COR.COEFF. FOR LOGNORMAL DIS. = 0.9841

COR.COEFF. FOR SMEMAX DIS. = 0.9821

COR.COEFF. FOR GUMBEL DIS. = 0.9788

ANNEXURE - II

a. Input file for solving Hooghoudt equation

```
2
1        0.4 5.0 75.0 0.0 50.0 75.0 150.0 350.0
2        0.4 5.0  75.0  0.0 50.0 75.0 150.0 500.0
```

b. Output file for hooghoudt equation

```
EXAMPLE 1
SPACING BY HOOGHOUDT EQUATION = 5388.54
EXAMPLE 2
SPACING BY HOOGHOUDT EQUATION = 5939.00
```

c. Input file for solving USBR equation

```
2
1        5.0 0.0 105.0 150.0 350.0 2.0 0.1 120.0
2        5.0 0.0  75.0 150.0 500.0 2.0 0.1 120.0
```

d. Output file for USBR equation

```
EXAMPLE 1
SPACING BY GLOVER-DUMM EQUATION = 3307.29
EXAMPLE 2
SPACING BY GLOVER-DUMM EQUATION = 2832.96
```

ANNEXURE – III

Threshold irrigation water quality, slope and EC_{50} of selected crops

Crop	Soil texture	Location	Piecewise linear model		
			EC_t	Slope (%)	EC_{50}
Wheat	Sandy loam	Agra	7.0	4.50	18.1
	Loamy Sand	Jodhpur	3.5	8.90	9.1
	Sandy loam	Karnal	7.9	8.70	13.6
	Dune Sand	Karnal	11.1	2.90	28.3
	Sandy loam	Hisar	3.4	5.50	12.5
	Sandy loam	Sampla	1.7	1.75	30.3
	Sandy loam	Kanpur	1.0	4.60	11.9
	Black clay	Dharwad	1.0	4.60	15.0
	Black clay	Indore	4.4	4.70	31.0
Barley	Sandy loam	Agra	8.3	2.20	23.5
	Sandy loam	Jobner	10.8	3.95	7.9
Rice	Sandy loam	Karnal	1.3	7.57	7.4
	Clay (S) (W)	Baptla	1.0 1.0	7.79 12.52	7.4 5.0
Maize	Black clay	Indore	1.0	7.17	8.0
	Black clay	Dharwad	1.0	2.84	18.6
Sorghum	Sandy loam (F) (G)	Agra Agra	3.2 6.2	2.96 3.82	20.1 19.3
	Black clay	Dharwad	1.0	5.80	9.6